SOME LIKE IT HOT

The Climate, Culture & Cuisine
of South Texas

A collection of recipes from the

☸ Junior League Of McAllen, Texas ☸

Library of Congress Catalog Card Number: 92-72326

ISBN: 0-9633359-0-1

Additional copies may be obtained by writing

Junior League of McAllen, Inc.

SOME LIKE IT HOT

P.O. Box 2465
McAllen, Texas 78502-2465
(210) 682-0743

| First Printing | November 1993 | 20,000 copies |
| Revised Edition | September 1995 | 20,000 copies |

Printed in the United States of America
TOOF COOKBOOK DIVISION

STARR ★ TOOF

670 South Cooper Street
Memphis, TN 38104

SOME LIKE IT HOT is a cookbook filled with a collection of recipes from the Junior League of McAllen, Inc. These recipes have been tested for flavor and quality. They are only a portion of the 1100 recipes submitted. We have gathered "Sun"-sational recipes to share with you, your family, and friends. We do not claim any of them are original. These recipes are "HOT" with flavor, not chili peppers! We hope that you will enjoy **SOME LIKE IT HOT** as much as we have enjoyed bringing it to you!

SOME LIKE IT HOT is dedicated to the outstanding traditions of the Junior League's volunteer work in McAllen, Texas. Since 1938, the Junior League of McAllen has raised countless dollars and contributed thousands of volunteer hours for community projects. Through our volunteer training, our members have become leaders throughout the Rio Grande Valley in the arts, in the education of children's issues, in health care, education, and in many other areas which serve to assure the best quality of life possible for all those who live in South Texas. We pledge to continue our commitment to community service through our voluntarism.

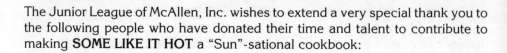

Acknowledgements

The Junior League of McAllen, Inc. wishes to extend a very special thank you to the following people who have donated their time and talent to contribute to making **SOME LIKE IT HOT** a "Sun"-sational cookbook:

Joe Michael Castellano   Cover and Dividers

South Texas Lifestyle Section Photographers

Steven Bentsen
Joe Michael Castellano
Robert Kracht
David K. Langford
John Scheiber
Chuck Snyder

South Texas Lifestyle Section photographs compliments of:

Cardenas, Whitis, Stephen, Corcoran, & McLain Law Firm
Hidalgo County Historical Museum
McAllen Chamber of Commerce
South Padre Island Visitor and Convention Bureau
Successful Attitudes Magazine
Texas Citrus Fiesta
Texas Sweet
The Club at Cimarron

Chef Mannie Garcia   Food Stylist

Historical references taken from: **A Shared Experience** by Mario Sanchez

The members of the Steering Committee of **SOME LIKE IT HOT** dedicate this cookbook to those whose commitment to excellence have made this cookbook a reality.

Junior League of McAllen, Inc. **1991-1992 President** Peggy Baldwin

Junior League of McAllen Volunteers

Chairman	Jo Ann Braun
Co-Chairman	Marsha Hart
Theme Development	Mary Ellen Stocker
	Sonia Perez
Color/Design	Sally Tankersley
Content Development	Deborah McClellan
	Kim Davis
Editing/Proofing	Beverly Stansberry
Computer Input	Mary Beth Devine
Marketing/P.R.	Jodi Hamer
Special P.R.	Jerry Talbott
DFM Coordinator	Patti Singer
Bookkeeper	Irma Garza

Sustaining Advisors

Jane McGilvray
Natalie West
Austin Miller

Janette McDonald
Pam Moore
Gloria Moore
Karen Moxley
Marion Wilson
Hilda Lewin
Jane Talbot
Carol Edrington
Glenda Ramee
Norma Cardenas
Jane Talbot

Recipe Collecting and Testing Committees

Appetizers/Beverages	Colleen Hook	Laurie Davenport
		Sarah Thomas
Soups/Salads	Shan Rankin	Bailey Gurwitz
		Elsie Hirsch
		Vicki Layer
Breads	Janice Fisher	Norma Cardenas
Entrees	Catherine Pons	Paula Hettler
Side Dishes	Laura Ilgun	Cissy Owens
		Jackie Stakes
Sweets	Ann Greenfield	Carolynn Nelson

Research and Copy for South Texas Lifestyle Section

Debbie Brooks
Jodi Hamer
Sonia Perez
Mary Ellen Stocker

The Junior League of McAllen, Inc. has contributed their volunteer time in order to bring a little Valley sunshine into your lives.

Vicki Aikman	Diane Farley	Tita Longoria	Amy Seitz
Kim Albracht	Debbie Fitch	Gloria Lowe	Sandra Semper
Diana Albrecht	Pat Fletcher	Laurie Lyde	Karen Shea
Elvia Arriaga	Shellie Foster	Carol Mackie	Ginger Simpson
Yvonne Bailes	Sheri Fox	Margot Madsen	Pamela Simpson
Raquel Baker	Bethann Freeland	Madhu Mangi	Vicki Skaggs
Yolanda Barrera	Cathy Freeland	Maree Mangum	Tricia Skalitsky
Gerry Batte	Jayne Freeland	Tina Martin	Lea Smith
Jane Beakey	Sharon Friedrichs	Janice Matthews	Melissa Smith
Edna Becerra	B-B Garza	Margie McCreery	Norma Somohano
Marge Bentsen	Velma Garza	Debbie McDaniel	Rosie Sotelo
Nell Bentsen	Becky Gerling	Beth McIntyre	Cherie Stiers
Beth Bergh	Dixie Goldsmith	Mary McKeever	Cathy Taylor
Pam Biddle	Bobby Goodrich	Madelaine McLelland	Anne Thaddeus
Sally Biel	Mariella Gorena	Kathy McManus	Debbie Thibodeaux
Lucille Bilbrey	Sally Hamlin	Laura Mesquita	Mary Carmen Thomas
Jeanne Blackburn	Katherine Hardwicke	Jan Miller	Sandra Thomas
Inalee Blackwell	Gay Hargis	Cynthia Mills	Sara Tippitt
Karen Boggus	Farideh Hatefi	Kathy Mills	Dorothy Toothaker
Brenda Bowman	Nancy Hawkins	Marilyn Moffitt	Linda Trevino
Elaine Bready	Vickie Hawkins	Ann Moore	Tracey Twenhafel
Libby Brown	Stephanie Haynes	Beverly Moore	Maria Elena Useda
Sandy Bryant	Linda Heard	Lynda Moore	Janet Vachar
Nancy Burns	Susan Mary Hellums	Page Moore	Patricia Valadez
Kimberly Camp	Laura Lee Hicks	Renee Morgan	Chris Van Burkleo
Dardanella Cardenas	Marianne Hilmy	Tracie Morrow	Patti Van Burkleo
Blanca Castilla	Dee Dee Hockema	Paula Moxley	Patsy Vela
Linda Cattaruzza	Kelly Hoffman	Karen Munoz	Edna Villarreal
Mary Chaffee	Sandra Hofland	Pam Myers	Cathy Voit
Vanessa Chang	Roberta Hohenstein	Dianne Noriega	Doris Wagner
Lisa Cofoid	Melissa Holland	Barbara O'Riley	Thelma Waite
Margaret Colley	Kathy Hook	Beth Pace	Betty Walker
Kathy Collins	Esther Jenkins	Val Peisen	Janet Wallace
Jane Cozad	Mel Johnstone	Brenda Perez	Karen Wallace
Tracy Crain	Elizabeth Jones	Ruthie Person	Nancy Welch
Baudelia Crane	Mary Vance Jones	Josie Picou	Martha Whetsel
Joanna Crane	Joy Judin	Louann Pollock	Linda Whitis
Billie Day	Phyllis Kendrick	Sharon Posluszny	Candie Wicker
Maxine Dean	Carolyn Kern	Sharon Rabinowitz	Geen Wilkins
Betty De Leon	Diane Kibbe	Julie Ramirez	Rosie Wilkinson
Denise Denham	Gayle King	Linda Ramirez	Becky Williams
Nancy Dillard	Wileen King	Lea Ann Ramsey-Fritz	Stephanie Wilson
Kelly Dizdar	Holly Kinsolving	Julianne Rankin	Betty Bentsen Winn
Lolly Durso	Julie Kittleman	Lezlie Richards	Dodie Wooldridge
Marie Dyke	Mary Kittleman	Shelly Risica	Jennifer Wright
Dawn Edwards	Ruth Kuhl	Lori Schlesinger	Melinda Wright
Willavae Etchison	Maggie La Grange	Debra Schneider	Anita Yee
Ruth Evans	Jennifer La Mantia	Frana Schrock	Michelle Zamora
Leslie Ewers	June Landrum	Marilyn Schultz	Sue Zipper
Patty Fallek	Teresa Lee	Debbie Schuster	
Doris Fair	J. Nevelyn Lloyd	Mary Jane Schwarz	

Table of Contents

❂ Junior League Of McAllen, Inc. ❂

The Junior League of McAllen, Inc. is a non-profit organization of women committed to promoting voluntarism and to improving the community through the effective action and leadership of trained volunteers. Its purpose is exclusively educational and charitable.

VOLUNTARISM: We support and promote voluntarism as an essential component of our society and will continue to take appropriate action to ensure the effectiveness of trained volunteers.
Rationale: The purpose of the Junior League of McAllen, Inc. is to promote voluntarism, to develop the potential of its members for voluntary participation in community affairs, and to demonstrate the effectiveness of trained volunteers.

THE FAMILY: We support efforts to increase communication among family members and to promote the positive spirit of the family environment.
Rationale: The Junior League of McAllen, Inc. wants to focus on the entire family, rather than just parts of it, because it recognizes that things which affect one family member, in reality, affect the entire group.

QUALITY OF LIFE: We are committed to ensuring that all our activities will contribute toward maintaining and/or improving the quality of life of its members and those it serves on an intellectual, physical, socio-economic, educational, cultural, and emotional level.
Rationale: The Junior League of McAllen, Inc. wants to assure that everything it does contributes in a positive manner to the growth of the area.

SUBSTANCE ABUSE: We support efforts to educate the community on the effect of substance and alcohol abuse through the support of preventive and rehabilitative programs.
Rationale: The Junior League of McAllen, Inc. recognizes that substance abuse is a major social problem with far-reaching physical, economic, and psychological ramifications.

Projects supported and sponsored by
The Junior League of McAllen
focus on

CHILDREN AND EDUCATION

Programs have included, but are not limited to the following:

Adult Literacy
Anti-Drug Puppets
Boys' and Girls' Club of McAllen
Cancer Awareness League
Comfort House
Food Bank
Girl Scouts of America
Health and Safety Puppets

Hidalgo County Foster Children
Horizon Youth Services
Hospice
McAllen International Museum
Mujeres Unidas/Women Together
Nursing Homes

Palmer Drug Abuse Program
Regional School for the Deaf
Scholarship Awards
Teen Court
Texas Girls' State
Women's Education & Employment Services
Youth Crisis Center

With the sun glistening and the palms swaying in the balmy breezes, a blend of ocean spray and orange blossoms fills the air. The vivid hues of Mexican sarapes dance before your eyes as the romance of the Spanish language whispers in the distance. The mariachi trumpets announce the beginning of yet another radiant sunset on the Rio Grande.

Good morning Rio Grande Valley.

PHOTO: CHUCK SNYDER / COMPLIMENTS: McALLEN CHAMBER

You have arrived in the heart of the Texas Tropics, the magic Rio Grande Valley, where the mornings begin with a cool dawn and the average temperature of 74 degrees allows you to participate in outdoor sporting events year-round. There are more than 21 golf courses in the Rio Grande Valley.

This seductive environment evolved from the days when hardy colonizers ventured into a land inhabited by Mexican settlers who began the ranching tradition long before modern irrigation made this land the lush agricultural Valley it has become.

This Valley of South Texas weaves together a collection of more than thirty-five communities originally settled shoulder-to-shoulder along the Texas-Mexico border. Here, residents, largely of Mexican and American descent, share their heritage with Mexican neighbors, who built the southern side of the Rio Grande with their own unique historical tradition.

An international center of trade, meshed with a richly fertile agricultural region, produced a lifestyle evolved from the days when the Rio Grande River truly was an expansive waterway and artery for commerce along the lower Texas-Mexico border. Commerce remains the engine driving the growth and progress in this region.

We'll start way out west, taking you through a land settled long before the English discovered their new world at Plymouth Rock. Originally colonized by Jose de Escandon, the current day Rio Grande Valley was part of a region that was the capital of New Spain. Strongly rooted in its past, the Valley surges forward to its future as a key center for Texas, the United States and international development.

PHOTO: CHUCK SNYDER / COMPLIMENTS: CARDENAS, WHITIS, STEPHEN, CORCORAN, & McLAIN

The old Southern Pacific train station.

Almost ready to ride.

PHOTO: DAVID K. LANGFORD

The far western section of our Valley, near Rio Grande City, evokes classic memories of days when the Spanish Viceroy ruled this vast brushland as far as the eye could see. Historic old buildings tell the story of the early days of colonization while the surrounding desert-like countryside holds memories of ancient native inhabitants, a land where Juan Cortina, a popular Mexican rebel, fought injustice with his raids through the Valley and points north.

Rio Grande City prides itself as home of the famous "La Borde House". Erected in 1897 as a home and business, this magnificently restored building today houses a hotel and is officially listed in the National Register of Historic Places. This beautiful treasure of frontier times transports you to the days of cattle barons and gun-barrel justice.

Traveling down the Rio Grande River in the steps of the Spanish explorers, visitors pass the Los Ebanos Ferry Crossing, an ancient ford used by Spanish explorers to haul salt from local salt mines for the King of Spain. One can almost still hear the hoofbeats of a later era when Texas Rangers chased cattle rustlers across the river.

The lush bougainvilla.

The ferry remains the country's only government-licensed, hand-pulled, international ferry. Travelers can cross on the ferry and lend a hand in pulling it across the river as the flowing waters lap against the hull. Mingle with other passengers from Mexico, Texas, the United States, Canada or who knows where!

Continuing further south, the desert gives way to a more lush, tropical landscape, part of which takes in Bentsen-Rio Grande Valley State Park. The park grew from part of an early Spanish land grant and is now home to an abundance of regional wildlife and vegetation, including flowering bougain-villas and hibiscus. Bird watchers and horticulturalists can be seen trekking through its endless animal trails.

A farmer's "field" day.

Twelve months of sunshine allow a year-round growing season for nearly every kind of crop and vegetable sold throughout the country. Oceans of sugar cane wave under the sun alongside fields that later bear snow white cotton. Visitors travel roadways lined with fields of crops planted in an alluvial soil that make this area the state's largest producer of agricultural products. Vegetables, both commonplace and exotic, grow in abundance.

"Sun"-sational produce.

PHOTO: JOHN SCHEIBER / COMPLIMENTS OF: McALLEN CHAMBER

South Texas Grapefruit.

Texas Citrus Fiesta Product Costume Pageant.

PHOTO: ROBERT KRACHT / COMPLIMENTS: TEXAS CITRUS FIESTA

Groves of citrus stretch to the horizon displaying trees filled with oranges of unequalled sweetness and the famous Ruby Red, Rio Red and Star grapefruits. Popular for their rose-colored meat, these delectable fruits, the pride of Valley citrus growers, can be sampled at roadside markets stocked with a myriad of delicacies.

All eyes turn to Mission for one of the most unique events in the state, the Texas Citrus Fiesta, a 60-year tradition paying tribute to a bountiful harvest of citrus. This fiesta, held the last week in January, fills the city with arts and crafts shows, citrus exhibits, parades and product shows. The Parade of Oranges features floats made of Valley-grown products.

A showcase fiesta event is the Product Costume Show, featuring exquisite costumes adorned completely with Valley-grown crops. Through years of competition and skills, the costumes have become intricate works of art using Valley citrus, fruits, vegetables, flowers and foliage. These costumes have been featured at the Kennedy Center in Washington D.C., as well as the Agricultural Hall of Fame in Kansas City, and have been highlighted in national magazines.

For history buffs, the city of Mission also boasts the winter home of Former Secretary of State William Jennings Bryan. He was lured to the most tropic-like area in his time. Walk the streets where one of America's greatest orators and politicians pondered the fate of the nation.

Buck fever for white tail deer.
PHOTO: © STEVE BENTSEN

Ask any avid American hunter about the Rio Grande Valley and he'll fill the conversation with tales of the whitetail deer, wild turkey, exotic javelina, quail and other wildlife that inhabit the region.

Consequently, hunters find a different kind of paradise as they flock to the Valley. Traditionally, the first two weekends of September attract those stalking the white-winged dove in this, the only part of the United States where the migratory bird can be hunted.

Next on the Texas Tropical Trail comes McAllen, the heart of the Rio Grande Valley, where those seeking more traditional cultural entertainment will find the McAllen International Museum, which features an outstanding permanent collection of Mexican folk art and masks. For kids of all ages, the museum also houses an Earth Science Gallery containing a long-term exhibition of dinosaur tracks, fossils and minerals, plus top quality educational programs and traveling exhibits.

For a truly spectacular historical perspective of old Texas, take a route slightly north to Edinburg for a visit to the Hidalgo County Historical Museum. Its exhibits, depicting the Valley's unique history, receive national recognition.

The elusive mourning dove.
PHOTO: © STEVE BENTSEN

Set in the Hidalgo County Jail built in 1910, the museum houses a turn-of-the-century hanging tower. Exhibits on Indian life and archeology, Spanish colonial settlements, the Mexican War, the Civil War, and Indian raiders attract thousands of visitors each year. Visitors are also captivated by displays of a time when steamboats were the life of the Rio Grande River. Articles and artifacts on early ranch life, the "Bandit Era", early law enforcement and early agriculture and land developers all uniquely preserve and present

McAllen International Museum Mexican Folk Art Display.

history for viewing. A tour of this museum truly represents a walk through several centuries of Valley life.

For the more intrepid, venture to a lake on the outskirts of Edinburg where local Indians once mined salt for the King of Spain. During the Civil War, this same site became part of a vital trade route for the confederacy through Mexico.

Hidalgo County Historical Museum Confederate Collection.

Continuing southeast, visitors come upon a "cowtown" atmosphere in the city of Mercedes. In 1939, the Rio Grande Valley Livestock Show and Rodeo originated as a place for ranchers to exhibit their livestock. Today, 4-H and FFA youngsters gather to show and sell their livestock. A full-fledged rodeo, carnival and continuously scheduled displays and ranching activities round out this five-day event.

Further east, in the Harlingen area, travel the same route as Spanish explorers did when they began their colonization of Texas for Spain in fear that the French would claim it as their own.

Today, Rio Fest, an annual celebration of the arts held in Harlingen, is labeled as the largest art show in South Texas. Rio Fest features continual concerts and hands-on activities with indoor and outdoor exhibits. Children can mosey into a make-believe "Western Town" where they can sidle up to the bar for a stiff lemonade or get locked up in the town calaboose.

Further down at the mouth of the Rio Grande River sit the sister cities of Brownsville, Texas, and Matamoros, Mexico. Just north of this population center lies the Palo Alto Battlefield, the field where the first two battles of the Mexican-American War were waged. Here, General Zachary Taylor inflicted heavy casualties on Mexican troops, a battle that eventually led to Texas' entry into the Union. The wind-swept site carries echoes of rifle shots, cannon barrages and the battle cries of hand-to-hand combat.

But the days of fighting are long gone, and today, when the people on both sides of the border get together, it is to celebrate the joining of their cultures. This unique blend of two cultures gives the Rio Grande Valley the best of two worlds.

Join a celebration of friendship and good will during "Charro Days" in Brownsville. A true bi-cultural gala with cuisine unique to our border gives visitors a delectable experience. Parades, mariachi bands, folkloric dancing and "grito" (yell) contests during Charro Days all help to foster close relations between the two countries.

COMPLIMENTS: HIDALGO COUNTY HISTORICAL MUSEUM

Do they jingle? Spanish spurs; circa 1700.

A Folkloric dancer.

Thousands of visitors know Brownsville as the location of the world famous Gladys Porter Zoo, which in turn is home to more than 1500 mammals, reptiles and birds. Animals live in areas created to reflect their natural homes in the wild. Opened in 1971, this beautiful, popular multimillion dollar facility enjoys international attention for its care and research on our vanishing wildlife.

Not too far from here, you'll spot the Point Isabel Lighthouse—one of the oldest functioning lighthouses on the Texas Gulf Coast. Climb up the steps to see how its shining beacon guided many a ship through the Gulf waters and averted many tragedies.

Across the bay, sun-worshippers head out to South Padre Island, often referred to as the "brightest star on the Texas coast." This part of the Texas coast includes a 34-mile barrier reef island at the southernmost tip of our Valley. Its harbor sheltered Confederate ships trying to run the Union blockade during the Civil War.

Gladys Porter Zoo.

PHOTO: JOE MICHAEL CASTELLANO

South Padre Island Hobie Cat Race.

Sand traps at The Club at Cimarron.

Last catch of the day.

*N*ow known for pristine beaches and its wealth of water sports, the island stands as a land of wind-carved dunes on miles of clean, white beach surrounded by the sparkling waters of the Gulf of Mexico and tranquil sunsets. Take a dip in the soothing gulf waves or relax on the soft sand under the Valley sun.

Sportsmen know the island as one of the best fishing spots in the Southwest with record-setting catches to prove it. Fishing is great anytime, but the island comes alive during the month of August when the Texas International Fishing Tournament (TIFT) attracts anglers from throughout the world. For more than 50 years, TIFT has proudly promoted the importance of coastal conservation efforts.

Designed with the entire family in mind, special children's activities, dinners and awards make this three-day fishing tournament a favorite regional gathering. From the tiniest amateurs to the seasoned angler, all ages find something at this event, even if it's just the great seafood in the area.

While the Valley's current inhabitants have progressed well beyond the days of the Texas Revolution and the Civil War, they cherish the rich history and culture that makes the Rio Grande Valley a special part of the country.

From colonial Texas to the most modern resorts, the Rio Grande Valley in South Texas is truly a place like no other. Its people are a unique breed who choose to share a very special way of life. A life they love, a lifestyle worth sharing...

Day's end.

PHOTO: DAVID K. LANGFORD

HOME ON THE RANGE — RIO GRANDE CITY STYLE

South Texas Bloody Mary Mix (pg. 69)

Squash and Green Chili Soup (pg. 89)

Venison Steak Fromage (pg. 154)

Baked Valley Onions (pg. 204)

Mexican Spoon Bread (pg. 40)

Pound Cake with Lemon Sauce (pg. 226)

TEXAS CITRUS FIESTA DESSERT PARTY

Mimosa Punch (pg. 72)

Sunshine Punch (pg. 76)

Strawberries with Orange Cream (pg. 253)

Orange Blossom Cheesecake (pg. 230)

Sour Lemon Bars (pg. 241)

Orange Chiffon Cake (pg. 225)

Cappuccino Caramels (pg. 234)

Candied Grapefruit Peel (pg. 234)

"SUN"-SATIONAL TEE-OFF

Spicy Coconut Chicken Bites (pg. 59)

Mushroom Nachos (pg. 32)

Cherry Tomatoes with Crabmeat (pg. 62)

Hors d'oeuvre Pie (pg. 68)

Campechano Shrimp Cocktail (pg. 33)

Spinach Dip (pg. 63)

Magnifico Margarita (pg. 35)

MERCEDES BLUE RIBBON BRUNCH

Sunshine Special (pg. 73)

Grapefruit Wedge Salad (pg. 104)

Brunch Casserole (pg. 146)

Ranch Style Sausage and Grits (pg. 149)

Texas-Size Biscuits (pg. 124)

Uncooked Orange Marmalade (pg. 265)

A CHARRO DAYS FIESTA

Sangrita (pg. 35)

Ensalada Especial (pg. 38)

Tortilla Soup (pg. 38)

Sour Cream Chicken Enchiladas (pg. 44)

Sopa de Fideo (Vermicelli) (pg. 51)

Frijoles Rancheros (pg. 50)

Kahlúa Mousse (pg. 55)

FISHING FOR COMPLIMENTS — PADRE STYLE

White Wine Lemonade (pg. 36)

Avocado Shrimp Spread (pg. 64)

Speckled Trout Ceviche (pg. 34)

Orange Roughy with Red Peppers (pg. 185)

Wild Rice (pg. 210)

Vegetable Medley (pg. 218)

Buttermilk Pralines (pg. 233)

A FALL FESTIVAL IN THE TEXAS TROPICS

Elegant Salad (pg. 94)

Quail in Orange Sauce (pg. 154)

Tomatoes Florentine (pg. 217)

Butternut Squash Soufflé (pg. 213)

Whole Wheat Butterhorns (pg. 119)

Death by Chocolate (pg. 248)

PICNIC BY THE RIO GRANDE RIVER

Pineapple Sangría (pg. 73)

Festive Cheese Ball (pg. 65)

Smoked Brisket (pg. 138)

Corn Salsa (pg. 267)

Spicy Potato Salad (pg. 100)

Herbed Garlic Bread (pg. 115)

Kahlúa and Praline Brownies (pg. 54)

Mexican Cuisine

MEXICAN CUISINE

APPETIZERS
- Avocado Dip 31
- Bean Dip 31
- Campechano Shrimp Cocktail 33
- Mexicali Avocado Dip 31
- Mushrom Nachos .. 32
- Pico de Gallo 33
- Queso Flameado (Flamed Cheese) 34
- South Texas Layered Dip 32
- Speckled Trout Ceviche 34
- Spinach Tortilla Bites 33

BEVERAGES
- Magnifico Margarita 35
- Mango Daiquiri 35
- Mexican Hot Chocolate 36
- Sangrita 35
- White Wine Lemonade (Sangria Blanco) . 36

SOUPS & SALADS
- Black Bean Salad .. 37
- Ensalada Especial . 38
- Hearty Mexican Soup 37

- Taco Salad 39
- Tortilla Soup 38

BREADS
- Flour Tortillas 39
- Mexican Spoon Bread 40

MAIN ENTREES
- Award Winning Beef Fajitas 40
- Calabazita con Pollo (Chicken With Squash) 43
- Chicken Chalupas . 44
- Chicken Enchiladas with Spinach Sauce 45
- Chicken Fajitas and Salsa 46
- Chicken Guisada ... 47
- Chili Relleno Casserole 42
- Envueltos de Pollo (Chicken Envueltos) 47
- Mexican Casserole 41
- Mexican Pork Chops 43
- Oven Chicken Olé 48
- Pollo con Chili Cream Sauce (Chicken with Chili Cream Sauce) 48
- Pollo con Naranja (Chicken with Oranges) 49
- Soft Tacos 41
- Sour Cream Chicken Enchiladas 44
- Taco Grande 42

SIDE DISHES
- Frijoles Rancheros . 50
- Green Chili Rice Casserole 50
- Ranch Beans a la Charra 49
- Sopa de Fideo (Vermicelli) 51
- Spanish Rice 51

SWEETS
- Flan 53
- Kahlúa and Praline Brownies 54
- Kahlúa Mousse 55
- Mexican Chocolate Roll 56
- Pan de Polvo 52
- Pineapple Empanadas 52
- Sopapillas 53

Anti-Drug Puppets

For over 14 years, our volunteers have visited the McAllen Independent Schools each year to teach thousands of fourth graders to "Just Say No" to drugs. Our message stresses the importance of saying "NO" and learning to resist negative peer pressure. After each performance, a McAllen Police Officer gives a follow-up presentation of the "8 WAYS TO SAY NO" and allows students to express their ideas in realistic role playing situations.

These recipes are pictured on the previous page.

AVOCADO DIP

Serves: 15-20

2 (8 oz.) cartons avocado dip
1 (16 oz.) carton sour cream
1 (16 oz.) jar picante sauce
1 (10 oz.) package Longhorn
  cheese, grated

2 bunches green onions,
  chopped (tops included)

13 x 9 casserole dish

Layer ingredients in serving dish, beginning with the avocado dip and ending with the green onions. Refrigerate. Serve with tortilla chips.

MEXICALI AVOCADO DIP

Serves: 10-12

1 (10 oz.) can chopped
  tomatoes and chilies, drained
1 envelope onion soup mix
2 (3 oz.) packages cream
  cheese, softened
3 avocados, mashed

1 tablespoon Worcestershire
  sauce
1 teaspoon lemon juice
⅛ teaspoon salt
1 package tostado chips

Combine all ingredients except tostado chips until well mixed. Serve with chips.

BEAN DIP

Yield: 2 quarts

4 cups dry pinto beans
2 cups chopped onions
3 cloves garlic, chopped fine
2 teaspoons ground cumin seed
6 tablespoons bacon drippings

6 tablespoons chili powder
1 cup margarine
½ pound sharp Cheddar cheese,
  grated
Dash hot sauce

Cook beans with ten cups of water, onion, garlic, cumin, and fat until soft, about 3 hours. Add chili powder and salt to taste, about 3 teaspoons. Do not have beans too soupy. While still warm, mash in the margarine, cheese, and hot sauce. Beat until smooth. Serve warm in a chafing dish with tostados or corn chips.

MUSHROOM NACHOS

Yield: 24 appetizers

24 large mushrooms
2 tablespoons olive oil
¼ cup butter, melted
2 jalapeño chilies, finely diced
(or to taste)

1 (7½ oz.) can or jar of salsa
ranchera
1 cup grated Monterey Jack
cheese

Preheat broiler. Cookie sheet

Lightly butter cookie sheet. Remove the stems from the mushrooms. Reserve the caps; chop the stems. Brush mushroom caps with melted butter; arrange on a baking sheet. In a skillet, sauté chopped mushroom stems in olive oil until all moisture is absorbed. Remove from heat; add jalapeños. Fill mushrooms with sautéed mixture. Top with grated cheese and a dab of salsa ranchera. Broil until cheese is melted.

SOUTH TEXAS LAYERED DIP

Serves: 10-12

1 (15 oz.) can refried beans
2 cups sour cream
1 (¼ oz.) package taco
seasoning
1 (8 oz.) jar picante sauce
4 ripe avocados, mashed
2 teaspoons lemon or lime juice
2 medium tomatoes, chopped

1 bunch green onions, sliced
with tops
1 (8 oz.) package Cheddar
cheese, grated
1 (4 oz.) can ripe olives, pitted
and sliced
Tortilla chips

2 quart casserole dish

Spread refried beans on bottom of casserole dish. Mix sour cream and taco seasoning; spread on top of beans. Spread with picante; mix avocados with lemon or lime juice and spread over picante; layer tomatoes, onions, cheese, and olives. Cover, chill, and serve with chips.

 For a great vegetarian dish, try spooning pico de gallo onto hot buttered tortillas. Roll tortillas and enjoy!

SPINACH TORTILLA BITES

Yield: 100-150 appetizers

2 (10 oz.) packages frozen chopped spinach, thawed and well-drained
6 green onions, chopped
1 package ranch style salad dressing mix
1 cup mayonnaise
1 jar bacon bits, not imitation
1 cup sour cream
10 large flour tortillas

Cookie sheet

Mix together spinach, onions, dressing mix, mayonnaise, bacon bits, and sour cream. Spread on tortillas; roll to enclose filling. Place on cookie sheet, seam side down, wrap in plastic wrap and chill overnight. Slice into bite-size pieces, about ½ inch. Let stand at room temperature for 15 minutes before serving.

CAMPECHANO SHRIMP COCKTAIL

Serves: 10-12

4-6 avocados, diced
½ pound (16-20) small shrimp
2-3 tomatoes, diced
½ medium onion, diced
1-2 serrano chilies, seeded and finely chopped (optional)
¼ bunch cilantro, stems removed and finely chopped
Juice of 2 limes
Salt and pepper to taste

Boil shrimp that have been cleaned and deveined; then chop. Mix shrimp and remaining ingredients; refrigerate until ready to serve. Avoid over stirring so avocados will not get mushy.

PICO DE GALLO

Serves: 4-6

5 ripe plum tomatoes, chopped
1 small onion, chopped
1 tablespoon minced fresh cilantro
½-1 serrano chili, minced with seeds (or to taste)
1 teaspoon fresh lime juice or vinegar
Salt to taste

Mix all ingredients; allow to stand 1 hour. Serve at room temperature.

SPECKLED TROUT CEVICHE

Serves: 10-20

2 pounds fresh speckled trout,
cut into 1 inch cubes
2 cups fresh lime juice
2 tablespoons salt, divided
2 cups tomato juice
2 medium tomatoes, chopped
1 medium onion, chopped
3 tablespoons finely chopped
fresh serrano peppers

4 tablespoons white wine
4 tablespoons olive oil
10 scallions, chopped
1 teaspoon oregano
¼ teaspoon soy sauce
½ teaspoon Worcestershire
sauce

Marinate cubed fish in lime juice and 1 teaspoon salt for 3-4 hours. Rinse in a colander. Mix remaining ingredients and add fish. Refrigerate overnight to allow flavors to blend. Serve cold with crackers or tostado chips. Great for summer-time appetizer.

QUESO FLAMEADO (FLAMED CHEESE)

Serves: 6

¼ pound chorizo (Mexican
sausage), casing removed
½ cup chopped onion
2 tomatoes, diced

⅓ cup picante sauce
4 cups (1 lb.) shredded Queso
Asadero (or Monterey Jack)

Preheat oven to 350°.   9 x 9 casserole dish

Crumble sausage and brown in small skillet; remove to paper towels with slotted spoon. Drain all but 1 teaspoon of the drippings; add onions and cook until tender. Return sausage; add tomatoes and picante sauce. Simmer 15 minutes or until most of liquid has evaporated. In casserole dish, arrange half of the cheese. Top with the sausage mixture and cover with remaining cheese. Bake about 10 minutes or until cheese melts. Serve with warm tortillas or tortilla chips.

One minced garlic clove may be added along with onion.

MAGNIFICO MARGARITA

Serves: 4

Ice
1 (6 oz.) can frozen limeade
  concentrate
3 ounces tequila
3 ounces Cointreau or Triple
  Sec

Salt
Freshly sliced limes or whole
green olives, optional

Blender

Fill blender with ice; add can of limeade and equal portions of tequila and liqueur. Blend to desired consistency. Dip the top of serving glass in the blended mixture; roll in salt before filling. When ready to serve, top with fresh lime slices or green olives.

1. Add a spoonful of sugar or sugar substitute to blender if you prefer a sweeter drink.

2. If blended mixture is too thick, add a little water.

3. If you have no Margarita glasses, champagne or sorbet glasses work well for serving.

SANGRITA

Serves: 10

1 (46 oz.) can tomato juice
6 tablespoons lime juice
6 tablespoons Worcestershire
  sauce
2 cups orange juice

Dash hot sauce
2 teaspoons onion juice
Salt and pepper
8 ounces tequila

Mix all ingredients and pour over ice ring.

MANGO DAIQUIRI

Serves: 2

1 cup mango flesh
3 large strawberries
½ cup light rum
  Juice of 2 limes

2 tablespoons superfine sugar
2 cups crushed ice
  Strawberries for garnish

In blender, blend all ingredients until smooth and frothy. Serve in chilled glasses.

WHITE WINE LEMONADE (SANGRÍA BLANCO)

Serves: 6

1 lemon
2 limes
2 oranges
½ cup water
½ cup sugar

Cracked ice
1 quart Chablis
1 cup sparkling mineral water
2 ounces orange liqueur

Cut a slice from the ends of the fruits to make them stand flat. Reserve fruit and place the end slices in a small saucepan with the water and sugar. Bring to a boil over medium high heat, stirring to dissolve sugar. Lower heat and simmer for 3 minutes to make a simple syrup. Remove from heat and let cool to room temperature. Strain the syrup, squeezing juices from the fruit. Discard the fruit end slices. Chill the syrup. May be kept one week.

On the day of serving, use a 2mm (thin) slicing disc to process the lemon and limes. Use a 3 or 4mm slicing disc to process the oranges. Discard seeds, chill the slices, covered, until ready to use. To serve the sangría, pour the syrup into a bowl or large pitcher filled with cracked ice. Add the reserved fruit slices and stir in the wine, mineral water, and liqueur. Serve immediately.

MEXICAN HOT CHOCOLATE

Serves: 4

4 ounces Mexican chocolate or
  3½ ounces semi-sweet
  chocolate plus 1 teaspoon
  cinnamon
4 tablespoons sugar

4 cups whole milk, scalded
Pinch salt
1 teaspoon vanilla, preferably
  Mexican
2 eggs

Break the chocolate into pieces. Place in food processor using the metal blade. Pulse several times to break up the pieces, then run machine continuously until finely chopped. Add the sugar through the feed tube with the motor running. With the motor still running, pour 1 cup of the scalded milk through feed tube. Add the vanilla, salt, and eggs in the same manner. Process until you have a good froth. Transfer to the remaining scalded milk in the saucepan, stirring constantly over low heat. Serve immediately.

HEARTY MEXICAN SOUP

Serves: 6-8

1 medium onion, chopped
1 (4 oz.) can chopped green chilies
1 jalapeño pepper, chopped (optional)
2 cloves garlic, minced
½ pound cooked chicken, diced
1 (14½ oz.) can tomatoes
1 (10 oz.) can tomatoes with green chilies
1 (14½ oz.) can beef broth
1 (14½ oz.) can chicken broth
1 (10¾ oz.) can tomato soup

2¼ cups water
1 teaspoon ground cumin
1 teaspoon chili powder
½ teaspoon lemon pepper seasoning
⅓-¼ teaspoon cayenne pepper
2 teaspoons Worcestershire sauce
4 soft corn tortillas, cut in 1 inch pieces
¼ cup grated Monterey Jack or Cheddar cheese
1-2 cubed avocados (optional)

5 quart stockpot

Drain juice from tomatoes into a skillet. Sauté onion, chilies, jalapeño, and garlic; transfer to stockpot. Add remaining ingredients, except tortillas, cheese, and avocado. Simmer uncovered 1 hour. Add tortillas; cook 10 minutes. Pour over cheese and avocado in bowls.

BLACK BEAN SALAD

Serves: 4-6

⅓ cup vegetable oil
¼ cup lime juice
4 tablespoons chopped cilantro
1 tablespoon minced pickled jalapeño
1 teaspoon minced garlic
½ teaspoon ground cumin
Dash of black pepper
Salt to taste

1 (16 oz.) can black beans, rinsed and drained
2 green onions, chopped
⅓ cup chopped yellow bell pepper
⅓ cup chopped red bell pepper
1 ripe tomato, coarsely chopped
1 avocado, sliced

Combine oil, lime juice, cilantro, jalapeño, garlic, cumin, salt, and pepper. Add beans and onions; toss well to coat. Cover and refrigerate several hours or overnight. When ready to serve, add peppers and tomatoes. Toss gently. Garnish with sliced avocado.

TORTILLA SOUP

Serves: 8

1½ medium onions, chopped
1 green bell pepper, chopped
2 carrots, sliced in thin rounds
¼ bunch celery, chopped
2 potatoes, chopped
3½ quarts chicken stock, heated
2 tablespoons chili powder
1 tablespoon comino (ground cumin)
1 teaspoon white pepper

1 teaspoon garlic powder
¾ cup garbanzo beans, cooked
1 pound chicken, cooked and chopped
2 tablespoons chopped fresh cilantro
Fresh lime, sliced
1 avocado, sliced
Fried tortilla strips

Sauté onion and bell pepper in small amount of oil for 60 seconds. Add carrots, celery, and potatoes and sauté for 2 minutes. Add hot chicken stock and seasonings; simmer for 5 to 10 minutes, or until vegetables are tender. Add garbanzo beans, chicken, and fresh cilantro. Garnish with fried tortilla strips, lime, and avocado.

ENSALADA ESPECIAL

Serves: 6-7

Salad:
1 head lettuce, torn in chunks
3 medium tomatoes, cut in small cubes
1 cup slivered onions
1 large cucumber, seeded and sliced
6 strips crisp bacon, crumbled

½ cup sliced radishes
8 ounces Cheddar cheese, grated
2 large avocados, sliced
1 (4¼ oz.) can sliced black olives
Slices of dill pickle to garnish

Combine all salad ingredients in a salad bowl.

Dressing:
6 ounces extra virgin olive oil
3 teaspoons Maggi seasoning
3 teaspoons Worcestershire sauce

1½ teaspoons dry mustard
8 tablespoons fresh lime juice
½ teaspoon celery salt

Combine all dressing ingredients; mix well. Pour over salad; toss gently.

TACO SALAD

Serves: 4

1 pound ground beef
½ cup chopped onion
1 can kidney beans, drained
1 head romaine lettuce, torn
  into bite-size pieces

2 tomatoes, diced
1 avocado, diced
1 (10 oz.) bag corn chips
1 bottle Green Goddess salad
  dressing

Brown ground beef and onion. Drain, add kidney beans and heat. Set aside. In large salad bowl, mix lettuce, tomato, avocado, and 1 cup of corn chips. Add meat mixture. Toss with dressing to taste. Serve with remaining corn chips on the side.

FLOUR TORTILLAS

Yield: 10 tortillas

2 cups flour
1 teaspoon salt
1 teaspoon baking powder

5 heaping tablespoons
  shortening
½ cup hot water

Blend flour, salt, and baking powder with a fork. Cut in shortening; add hot water. Complete mixing the dough by hand and shape into large ball. Let rest 5-10 minutes. Divide dough into 10 portions and shape into balls. Lightly flour surface if dough is sticky. Roll each ball into a thin circle. Cook on a griddle over medium heat, approximately l minute; then turn and cook about 30 seconds on other side. Do not overcook as the tortillas will get too hard!

This recipe requires a coarse flour. We recommend Pillsbury flour.

 To keep tortillas hot at the table, serve between 2 heated plates.

MEXICAN SPOON BREAD

Serves: 6-8

1 cup yellow corn meal
2 eggs
1 (16 oz.) can creamed corn
½ cup sweetened condensed
    milk
⅓ cup oil

½ tablespoon baking soda
1 (8 oz.) can whole green
    chilies
12 ounces cheddar cheese
    grated

Preheat oven to 350°.
11 x 9 baking pan

Lightly grease baking pan. Thoroughly mix corn meal, eggs, corn, sweet milk, oil, and baking soda in a large bowl. Pour ½ of the batter into prepared baking pan. Rinse green chilies and remove the seeds. Lay chilies on top of the batter and sprinkle with cheese. Pour remaining batter atop cheese. Bake 40 to 45 minutes.

Variation: Substitute sweetened condensed milk with evaporated milk and green chiles with 2-3 jalapeños for a caliente alternative.

AWARD WINNING BEEF FAJITAS

Serves: 10-12

6 pounds fajitas, well-skinned
10-12 fresh limes
⅓ cup teriyaki sauce
Fajita seasoning with
    tenderizer, to taste

California garlic salt (blended
    with parsley and other
    spices), to taste
2 tablespoons olive oil

Four hours prior to cooking, squeeze fresh lime juice on both sides of meat and sprinkle lightly with teriyaki sauce. Let stand 2-2½ hours. Drain excess liquid and sprinkle garlic salt and fajita seasoning over meat. Using a small brush, dab olive oil all over both sides of the meat. This will prevent burning the meat over hot flames. Cook marinated fajitas quickly over very hot coals. Flame cook until darkened, turn meat over and repeat. Remove from flames once both sides are darkened. Cut into 1-inch strips across the grain. Serve hot with tortillas, pico de gallo, and margaritas.

MEXICAN CASSEROLE

Serves: 6-8

2 pounds lean ground meat
1 large onion, chopped
1 pound grated mild Cheddar
cheese, divided
1 (15 oz.) can Ranch Style
beans, undrained
1 tablespoon chili powder

1 teaspoon garlic powder
Salt and pepper to taste
12 corn tortillas
1 (10½ oz.) can condensed
cream of chicken soup
1 (10 oz.) can diced tomatoes
and green chilies

Preheat oven to 325°.   13 x 9 casserole dish

Lightly grease casserole dish. Sauté onion and meat until meat is browned. Drain. Add undrained beans, chili powder, garlic powder, salt, and pepper. Mix well. Cut 6 tortillas in strips; line bottom of dish. Cover with meat mixture, then sprinkle ½ of cheese on top. Cut other 6 tortillas in strips and place on top, then top with remaining cheese. Mix condensed soup with tomatoes and green chilies and pour over top. Bake uncovered for 1 hour.

SOFT TACOS

Serves: 4-6

1 pound hamburger meat,
browned
½ onion, chopped
8 ounces Velveeta cheese
1 (8 oz.) can tomato sauce

1 (11 oz.) can evaporated milk
1 (10 oz.) can diced tomatoes
and green chilies
⅓ cup vegetable oil
1 package (12) corn tortillas

Preheat oven to 350°.   13 x 9 baking pan

Cook meat and onions until meat loses red color. In a saucepan, melt the Velveeta cheese with the tomato sauce, milk, and tomatoes and green chilies. Pour ⅓ cup of the cheese mixture into the meat, saving the rest. Heat the vegetable oil in a small frying pan. Dip the tortillas into the heated oil until slightly brown on the edges. Be careful not to over-brown. Place a small portion of the cooked meat into each tortilla and roll it up. Place the rolled tortilla in pan. Repeat with all the remaining tortillas. When the pan is full, pour the rest of the cheese sauce over all the rolled tortillas. Heat for 5-10 minutes or until the tacos are hot but not brown.

TACO GRANDE

Serves: 6-8

2 pounds ground beef
1 package taco seasoning mix
1½ cups water
1 (16 oz.) package tostados
1 cup grated Cheddar cheese
1-2 small onions, chopped

10 ripe black olives, chopped
1 tomato, chopped
1 cup sour cream
1 cup shredded lettuce
1 cup guacamole or avocado dip

Brown ground beef in large skillet. Drain excess fat; add taco seasoning mix and water. Bring to boil; reduce heat; simmer for 25 minutes, stirring occasionally. Put tostados in bowl and remaining ingredients in individual bowls around table. Serve meat in pan. To make dish start with tostados, then a spoonful of hot meat, cheese, onions, olives, tomato, sour cream, lettuce, and guacamole.

CHILI RELLENO CASSEROLE

Serves: 8

½ cup butter or margarine
1 (7 oz.) can whole green
   chilies, rinsed and drained
2 cups shredded Monterey Jack
   cheese

¾ cup shredded Cheddar cheese
3 eggs, beaten
¾ teaspoon salt
1 cup Bisquick
2 cups milk

Preheat oven to 350°.   13 x 9 baking pan

Melt butter in pan. Cut chilies into 1 inch wide strips; lay evenly over melted butter. Top chilies with combined cheeses. Mix together beaten eggs, salt, Bisquick, and milk; pour over cheese. Bake for 45 minutes until lightly browned.

 Add a small amount of lemon or lime juice to prevent seeded avocados from turning brown.

MEXICAN PORK CHOPS

Serves: 4-6

6 thick pork chops
Salt and pepper to taste
¾ cup long grain rice
1 (13½ oz.) can tomato sauce
1¾ cups water

2 tablespoons taco seasoning
mix
1½ cups shredded Cheddar
cheese

Preheat oven to 350°.

13 x 9 baking pan

In large skillet, brown pork chops; season with salt and pepper. Place in single layer in pan. Pour rice over the chops. Mix tomato sauce, water, and taco seasoning. Carefully pour over chops and rice. Cover with foil; bake for 1 hour. Uncover; cover with cheese; bake 15 minutes longer.

CALABAZITA CON POLLO
(CHICKEN WITH SQUASH)

Serves: 8

1 chicken
Salt and pepper to taste
2 tablespoons butter
4-6 medium sized squash
4 tomatoes, peeled
2 ears of corn
3 cloves garlic, minced

1-2 large onions, sliced
4 jalapeños, seeded
2 green bell peppers, chopped
1 pimiento, diced
1 can beer (optional)
2 tablespoons butter, melted
2 tablespoons flour

Preheat oven to 325°.

In baking pan, place 1 cup water and chicken, liberally seasoned with salt, pepper, and butter. Bake, covered, until tender. Skin and bone chicken, reserving broth. Wash and cut vegetables into bite-sized pieces. Remove corn from cob; add garlic and onion. Cut jalapeños into half-inch squares. Place cooked chicken, broth, and vegetables in large saucepan. Add beer or 1½ cups water. Cover and simmer 15 to 20 minutes over low to medium heat. Do not let ingredients stick to bottom of pan. Mix butter with flour and add just enough to thicken mixture. Serve in bowls with flour tortillas.

CHICKEN CHALUPAS

Serves: 6-8

6 chicken breast halves, skinned and boned
1 cup chopped green onions, tops included
2 large jalapeños, seeded and chopped
¼ cup jalapeño juice
1½ cups sour cream
1 jar pimientos
Fried chalupa shells
8 ounces Monterey Jack cheese, grated
Shredded lettuce
Diced tomato
Diced onion
Guacamole

Boil chicken and chop into chunks. Mix chicken with onions, jalapeños, juice, pimientos, and sour cream. Spread over chalupa shells; sprinkle with grated cheese. Broil until bubbly; top with lettuce, tomato, onion, and guacamole.

SOUR CREAM CHICKEN ENCHILADAS

Serves: 5

4 chicken breast halves, skinned and boned
4 tablespoons margarine
1½ cups sour cream
1 (10¾ oz.) can cream of mushroom soup
½ can (10 oz.) diced tomatoes and green chilies with juice
1 (7¼ oz.) can sliced black olives, drained
8 ounces Monterey Jack cheese, shredded
1 medium onion, grated
10 flour tortillas

Preheat oven to 350°.   13 x 9 baking pan

Lightly grease baking pan. Simmer chicken breasts in lightly salted water until tender, 20 minutes, and shred. Set aside. In a large skillet melt margarine. Stir in sour cream, mushroom soup, and tomatoes and green chilies. Set aside ½ cup of the cheese and 1 cup of the sour cream mixture. Dip each tortilla in the sour cream mixture; fill with ¹⁄₁₀ of the chicken, cheese, onions, and olives. Roll and place in pan. Repeat with the remaining 9 tortillas. Top with cheese and sour cream mixture. Bake covered for 45 minutes. Uncover and brown slightly until bubbly.

Serve with a green salad with avocado dressing. This dish may be frozen for up to one month. If refrigerated or frozen, allow extra baking time.

CHICKEN ENCHILADAS WITH SPINACH SAUCE

Serves: 8-12

2 dozen corn tortillas
1-1½ cups vegetable oil
3 cups shredded cheese
(Cheddar or Monterey Jack)

1 medium onion, chopped
1 (3-4 lb.) chicken, boiled,
skinned, boned, and shredded

Preheat oven to 350°.   2 (13 x 9) baking pans

Lightly grease baking pans. Dip tortillas one at a time in very hot oil just long enough to slightly darken and soften. Generously spread a one-inch central strip of each tortilla with shredded cheese, chopped onion, and shredded chicken. Roll up, using toothpicks to secure if necessary. Place seam side down, side by side in baking dishes. Try to leave at least ½ inch clearance between enchiladas and the sides of the pan; do not pack the enchiladas too tightly. This will allow room for plenty of sauce to coat and penetrate the enchiladas. Cover generously with spinach sauce.

Creamy Spinach Sauce:

1 (10 oz.) box frozen chopped
   spinach
1½ cups water
2 (10½ oz.) cans condensed
   cream of chicken soup
3-4 green onions, minced

2 (4 oz.) cans chopped green
   chilies, undrained
¼ teaspoon salt
1 pint sour cream
1½ cups shredded cheese (same
   type as in enchiladas)

Cook spinach in a large saucepan with water. Add condensed soup, green onions, chilies, and salt. Stir with a large whisk or put in a food processor to blend smoothly. After mixture simmers, remove from heat; add sour cream. Mix well. Pour over enchiladas. Sprinkle with shredded cheese. Bake uncovered for 30 minutes. Remove toothpicks before serving or storing.

This is the perfect do-ahead recipe, since flavor is enhanced with time! Just bake, refrigerate, and reheat before serving.

 To heat corn tortillas, place tortillas in an unsealed plastic bag; heat in microwave in 20-second increments until tortillas are heated.

CHICKEN FAJITAS AND SALSA
Serves: 12

2 pounds chicken breasts, boned and skinned
2 medium onions, thinly sliced
½ cup lime juice
4 teaspoons finely chopped garlic
½ teaspoon freshly ground pepper
1½ teaspoons ground cumin
1 teaspoon dried oregano, crushed
1 red bell pepper, cut into thin strips
1 green bell pepper, cut into thin strips
Salsa
Fresh cilantro
1 dozen tortillas

13 x 9 baking pan

Cut chicken breasts into strips. Place chicken and onions in pan. Set aside. Mix together lime juice, garlic, pepper, cumin, and oregano. Pour over chicken and onions; cover; marinate 2-4 hours in refrigerator, stirring occasionally. (Do not exceed 4 hours or the texture of the chicken will be altered.) Heat skillet; add chicken and onions, red and green peppers, and 1-2 tablespoons marinade (to keep mixture from sticking to the pan.) Stir fry just until chicken becomes opaque and is no longer pink. Do not overcook. Place about ½ cup of mixture in the center of a warm tortilla. Top with salsa and cilantro. Accompany with Spanish Rice.

Salsa:

1 pound tomatoes, diced
¾ cup diced onion
2 tablespoons chopped cilantro
½ jalapeño pepper, seeded and chopped
½ teaspoon chopped garlic
¾ teaspoon ground cumin
¾ teaspoon dried, crushed oregano
⅛ teaspoon salt
1 tablespoon freshly squeezed lemon juice
1 tablespoon freshly squeezed lime juice

Combine all ingredients. Cover and refrigerate for at least 2 hours before serving.

 For chicken meat that will be more tender and less stringy, use a pair of kitchen scissors to cut up the cooked, boned chicken.

CHICKEN GUISADA
Serves: 6-8

2 pound fryer
3 cloves garlic, peeled
2 teaspoons salt
¼ cup oil
1 teaspoon pepper

3 tomatoes, chopped
1 onion, chopped
1 serrano pepper, chopped
3 large bay leaves, crushed
Grated cheese

Place chicken, garlic, and salt in pot; cover with water and boil until tender. Skin, bone, and shred chicken and place in skillet with oil; sauté. Add remaining ingredients; simmer about 10 minutes, covered. Can be served in flour tortillas topped with grated cheese or as a meat dish with Spanish Rice and Frijoles Rancheros.

ENVUELTOS DE POLLO (CHICKEN ENVUELTOS)
Serves: 4-6

1 (2-3 pound) fryer
6 cups water
1 tablespoon salt
½ small onion, diced
1 tablespoon cooking oil
2 (15½ oz.) cans whole
   tomatoes

1 clove garlic
1 teaspoon mixed spices
   (peppercorns and cumin)
2 cups chicken broth
12 corn tortillas

Preheat oven to 350°.   13 x 9 baking pan
Lightly grease baking pan. Boil cut-up chicken in salted water. Cover, simmer for 40-50 minutes. Reserve broth. Remove skin and bones from chicken. Dice chicken; set aside. In skillet, sauté onion in oil. Grind garlic and mixed spices, add a little water and add to sautéed onions. Simmer 1 minute. Stir in tomatoes and broth. Simmer 5-10 minutes. Dip tortillas one at a time in tomato mixture; place in pan. Fill with chicken; roll like an enchilada. Place seam-side down to keep filling in place. Pour remaining sauce over enchiladas. Bake for 10 minutes. Serve hot.

OVEN CHICKEN OLÉ

Serves: 4

3 chicken breasts, boned and
  halved
3 ounces cream cheese
1 cup mild picante sauce

1 teaspoon ground cumin
½ cup sliced green onions
Crushed tortilla chips (not
  Fritos)

Bake at 350°.

13 x 9 baking pan

Pound chicken breasts to flatten. Mix together cream cheese, picante sauce, ground cumin, and sliced green onions. Place a spoonful of mixture on each piece of chicken breast and fold over. Pour remaining sauce over top of chicken rolls. Spread crushed tortilla chips over sauce. Bake for 45 minutes, uncovered.

POLLO CON CHILI CREAM SAUCE

Serves: 6

2 fresh Anaheim (also known as
  California) chilies
1½ cups whipping cream
1 cup thinly sliced red onion
⅛ cup chopped fresh cilantro
1 clove garlic, minced (or ¼
  teaspoon garlic powder)

6 boneless chicken breast
  halves
½ teaspoon salt
⅛ teaspoon pepper
¼ teaspoon garlic powder
1 large tomato, peeled, seeded,
  and chopped

Prepare barbeque grill.

Char chilies over gas flame or in broiler until blackened on all sides. Wrap in a paper bag and let stand 10 minutes to steam. Peel, seed, and chop chilies. Combine chilies, cream, onion, cilantro, and garlic in a large saucepan over medium-high heat. Boil this mixture until reduced to a thick sauce, stirring occasionally, about 5 minutes. Season chicken breasts with salt, pepper, and garlic powder. Place chicken on prepared barbeque grill and cook until springy to the touch, about 5 minutes per side. Cut chicken diagonally into thin slices. Add tomato to sauce and simmer. Season sauce with salt and pepper. Spoon sauce onto plates. Place chicken atop sauce and serve.

POLLO CON NARANJA (CHICKEN WITH ORANGES)

Serves: 4-6

1 4-pound broiler fryer chicken,
   cut in pieces
½ teaspoon salt
⅛ teaspoon pepper
⅛ teaspoon cinnamon
⅛ teaspoon ground cloves
3 tablespoons salad oil
2 whole cloves garlic
1 medium onion, chopped
1 cup orange juice

1 cup water
⅛ teaspoon saffron (optional)
2 tablespoons seedless raisins
   (optional)
1 tablespoon capers (or
   nasturtium seeds)
½ cup coarsely chopped
   almonds
3 Valencia oranges, peeled and
   sliced

Sprinkle chicken with salt, pepper, cinnamon, and cloves. In a large skillet, brown the chicken in oil over moderately high heat. Add garlic and onion; continue cooking until chicken is brown but not crusted. Add orange juice, water, saffron, raisins, and capers. Cover and cook over low heat until chicken is tender, about 40 minutes. Add almonds 5 minutes before serving. Remove garlic. Decorate with orange slices.

RANCH BEANS A LA CHARRA

Serves: 8

4 slices bacon, chopped
1 large onion, chopped
2 medium tomatoes, chopped
2 (23 oz.) cans Ranch Style
   beans

¼ cup chopped cilantro
1 (23 oz.) can water

Lightly brown bacon pieces, remove from skillet; sauté onions and tomatoes in drippings. Combine bacon, sautéed vegetables, beans, cilantro, and water. Simmer for 30 minutes.

You may use beans with jalapeños if desired.

FRIJOLES RANCHEROS

Serves: 20-2!

2 pounds dry pinto beans
¾ pound salt pork, ham, or
bacon
1 clove garlic, minced
1 tablespoon salt
3 tomatoes, chopped

1 medium onion, chopped
1 serrano chili, chopped
½ bunch cilantro, chopped
3 tablespoons thick and chunky
picante sauce

Place beans in large pot filled half full of water. Cook on medium low heat for
hour; add salt pork, ham or bacon, garlic, and salt. Cook for 2 hours on mediun
heat. Reduce heat to very low and cook 1 more hour. Refill pot with water an
add tomatoes, onion, chili, cilantro, and picante sauce. Simmer until beans ar
soft, adding water every 15 minutes if needed.

Do not stir beans too much or the soup will be very thick.

GREEN CHILI RICE CASSEROLE

Serves: ?

3 cups cooked white rice
1 cup sour cream
1 (10¾ oz.) can cream of celery
soup
⅛ teaspoon salt

1 cup grated Cheddar cheese
4 tablespoons butter, melted
1 (4 oz.) can chopped mild
green chili peppers

Preheat oven to 350°.

13 x 9 baking par

Lightly grease pan. Combine all seven ingredients. Pour combined ingredient:
into pan. Cover and bake for 30 minutes.

*When handling fresh chilies, wear rubber gloves, which protect not only your
hands, but also your face, eyes, and other sensitive areas with which your
hands come in contact. Always wash your hands and all utensils thoroughly
with hot soapy water after working with chilies.*

SPANISH RICE

Serves: 6

3 tablespoons oil
1 cup rice
Few slices of onion
Few slices of green bell pepper
½ can whole tomatoes with green chilies
½ teaspoon salt
½ teaspoon ground pepper
½ teaspoon garlic powder
½ teaspoon ground cumin
1 teaspoon chicken bouillon granules
1 cup hot water
½ cup canned or frozen peas and carrots (optional)

Heat oil on medium heat in skillet. Add rice, onion, and bell pepper and brown. Do not let rice burn! Blend tomatoes and spices in blender. It should measure 1 cup liquid. Add hot water and blend again. Add liquid mixture to rice and stir. Return to boil for 30 seconds. If desired, add peas and carrots at this time and stir once or twice only. Cover skillet with a tight fitting lid. Reduce heat to low and cook for 15 minutes.

SOPA DE FIDEO (VERMICELLI)

Serves: 6-8

2 cloves garlic
½ teaspoon cumin
½ teaspoon peppercorns
⅓ cup cooking oil
1 (10 oz.) package vermicelli
1 (14½ oz.) can whole tomatoes
¼ cup chopped green bell pepper
Salt to taste
3 cups boiling water

Grind garlic and spices. Add a little water; set aside. Break up vermicelli. In skillet on low heat, add oil and vermicelli; fry until golden brown. Add spices, tomatoes, bell pepper, salt, and boiling water; simmer 15-20 minutes.

1½ cups white rice may be substituted for vermicelli.

PINEAPPLE EMPANADAS

Yield: 3 dozen

4 cups flour
1 teaspoon sugar
¼ teaspoon salt
1¾ cups shortening
1 egg, well beaten
1 tablespoon white vinegar

½ cup water
1 (10 oz.) jar pineapple preserves
1 cup sugar
1 tablespoon cinnamon

Preheat oven to 400°.   Ungreased cookie sheet

Sift flour; mix with sugar and salt. Cut in shortening using 2 forks. In a large bowl combine well beaten egg, vinegar, and water. Using your hands mix the dough to form a ball. Chill for 30 minutes. Divide the dough into balls about 2 inches in diameter. With rolling pin, roll into circles about 2 inches in diameter. Place ½ teaspoon pineapple preserves on 1 side of each circle; fold the other half of dough over preserves; pinch edges to seal. Bake for 15-18 minutes.

While the empanadas are baking, mix sugar and cinnamon in a large bowl and set aside. Gently remove baked empanadas from cookie sheet and coat with cinnamon and sugar mixture. Place on platter to cool.

Other flavors of preserves may be substituted if desired.

PAN DE POLVO

Yield: 5-6 dozen

2 pounds flour (use premeasured size)
1 cup sugar
1 pound shortening (use premeasured size)

1 cup finely chopped pecans
1 egg (add an extra egg if mixture appears too dry)

Preheat oven to 350°.   Ungreased cookie sheet

Combine flour and sugar. Using a pastry blender, cut in shortening; add pecans. Beat egg and add to flour mixture. Knead in bowl and separate into 6 balls. Work each ball individually. Roll to ¼ inch thickness and cut with tiny cookie cutters. Bake 10-12 minutes (cookies will look pale). Roll in cinnamon sugar mixture.

SOPAPILLAS

Yield: 12

4 cups flour, sifted
4 teaspoons baking powder
1 teaspoon salt
1 tablespoon shortening
1 egg, beaten

1 cup water
Honey, cinnamon-sugar, or
powdered sugar for topping

Deep-fat Fryer

Mix together all dry ingredients; cut shortening in using a pastry knife. Add egg and water; mix well. Knead dough until smooth. Divide dough into four parts. Roll dough to thickness of ⅛th inch and cut into 3-inch squares. Deep-fat fry until golden brown. Top with honey, cinnamon-sugar, or powdered sugar.

FLAN

Serves: 8

1½ cups sugar, divided
6 whole eggs
1 (12 oz.) can evaporated milk
1 (14 oz.) can sweetened
   condensed milk

14 ounces fresh milk
2 tablespoons vanilla extract

Preheat oven to 350°.   2 quart mold

Caramelize ½ cup sugar in a small heavy skillet. Pour into mold and turn the mold around quickly, tilting it from side to side until the mold is coated on the bottom and halfway up the sides. Set aside. Beat eggs lightly; slowly add 1 cup sugar and beat well until all is dissolved in the eggs. Add milks and vanilla; beat about 2 minutes. Pour custard mixture into mold; set the mold in a hot water bath that comes about halfway up the side of the mold. Place on lowest rack of oven and bake for 1 hour and 20 minutes. Test by inserting a knife or toothpick in center. Remove when it comes out clean. Let cool about 1 hour. Invert pan onto a plate with a small ridge so that the caramel will not spill.

KAHLÚA AND PRALINE BROWNIES   Yield: 3 dozen squares

Praline crust:

⅓ cup light brown sugar, firmly packed
⅓ cup butter

⅔ cup sifted flour
½ cup finely chopped pecans

Mix together sugar, butter, flour, and pecans. Pat evenly over bottom of a 9 inch square pan and set aside.

Brownie filling:

2 (1 oz.) squares unsweetened chocolate
¼ cup shortening
¼ cup butter
2 large eggs
½ cup granulated sugar
½ cup light brown sugar, firmly packed

1 teaspoon vanilla extract
¼ cup Kahlúa
½ cup sifted flour
¼ teaspoon salt
½ cup chopped pecans

Preheat oven to 350°.

Prepare filling by melting chocolate with shortening and butter over low heat. Beat eggs with sugars and vanilla until blended. Stir in cooled chocolate mixture, then Kahlúa. Add flour and salt, mixing to a smooth batter. Stir in pecans. Pour into pan lined with praline crust and bake for 25 minutes or until done, being careful not to overbake. Let cool.

Kahlúa Butter Cream Frosting:

2 tablespoons butter, softened
2 cups sifted powdered sugar

1 tablespoon Kahlúa
1 tablespoon cream

Beat all ingredients until smooth and creamy. If necessary, beat in additional Kahlúa for good spreading consistency. Spread frosting over top of brownies; place in refrigerator ½ hour to set.

(Continued on next page)

(Kahlúa and Praline Brownies, continued from previous page)

Chocolate Glaze:

2 (1 oz.) squares semisweet
chocolate

1 (1 oz.) square unsweetened
chocolate

2 teaspoons shortening

Melt ingredients together over low heat. Stir to blend and cool before spreading evenly over frosting and cool until set. Cut into small squares.

KAHLÚA MOUSSE

Serves: 8

½ cup sugar
½ cup water
2 eggs
⅛ teaspoon salt
1 (6 oz.) package chocolate
chips

2 tablespoons cognac
3 tablespoons Kahlúa
1 pint heavy cream, whipped
Toasted almonds, slivered
Chocolate curls

Heat sugar in water until dissolved; set aside. Mix eggs, salt, and chocolate in blender. Add sugar mixture slowly and blend until thick. Cool. Add cognac and Kahlúa. Fold in half of the cream into chocolate mixture. Chill several hours. Serve in small wine glasses, top with remaining whipped cream, almonds, and chocolate curls.

MEXICAN CHOCOLATE ROLL

Serves: 12-14

½ cup dark seedless raisins, optional
½ cup chopped pecans, optional
¼ cup heated Kahlúa
8 large eggs, separated
½ cup sugar
9 ounces Mexican chocolate or 8 ounces German sweet chocolate plus ¼ teaspoon cinnamon

1 ounce unsweetened chocolate*
¼ cup boiling water
Pinch salt
Pinch cornstarch
2 cups heavy cream, chilled
⅓ cup powdered sugar
Additional powdered sugar
Unsweetened cocoa

Preheat oven to 350°.   17x11 inch jelly roll pan

Butter pan and line with buttered wax paper having the paper overlap the edges of the pan. Soak the raisins and nuts in the warm Kahlúa for 2 hours.

Cake: Beat egg yolks with the sugar until very thick. Set aside. Break the chocolate into 8 pieces; in food processor, finely grind chocolate. Pour the boiling water through the feed tube to melt the chocolate. Cool the chocolate by placing the workbowl in the freezer for about 10 minutes. Do not solidify. Return workbowl to machine base, add the beaten egg yolk mixture and pulse once or twice to combine. Leave in workbowl. Using a clean mixer bowl and beaters, beat the egg whites, salt, and cornstarch to stiff peaks. Pulse several spoonfuls of the egg whites into the chocolate mixture to lighten, then fold in the remainder by hand. Spread the batter in the prepared pan and bake 10 minutes. Lower temperature to 325° and bake an additional 10-15 minutes until cake is firm, but not dry or browned. Remove from oven and cover immediately with a cool, slightly dampened towel. Refrigerate at least 1 hour.

Filling: Whip the chilled cream until thickened. Then add the powdered sugar. Remove about ½ cup of the whipped cream and place in a pastry bag fitted with a star tip and reserve to garnish the dessert. Drain the raisins and nuts and fold into the remaining cream.

To assemble the cake roll: Remove the cloth from the cake, dust lightly with cocoa and invert onto a piece of wax paper placed on a work surface. Use the overlapping edges of waxed paper in the pan to loosen the cake. Peel off the wax paper and dust the cake with cocoa. Spread the Kahlúa flavored cream over the cake, leaving about 1½ inches on the long sides free of filling. Roll up from the long side, using the wax paper as an aid. Transfer, using the wax paper to lift, and place seam side down on a serving platter. Decorate the cake roll with rosettes of whipped cream and dust the entire surface with additional powdered sugar and cocoa. Chill until ready to serve. *If using La Fonda or La Popular brand chocolate, you will need to add the unsweetened chocolate to the recipe.

Appetizers
& Beverages

Refer to Mexican Cuisine section for additional recipes in this category.

Palmer Drug Abuse Program (P.D.A.P.)
Through the "Special People Program" our volunteers help children between the ages of 5-13 who have been affected by the chemical addiction of a family member. Our goal is to teach the children how to cope with their feelings and to value themselves by teaching them that they are truly "Special People." We have raised monies to distribute the "Special People" program literature throughout the schools and to expand the ever growing needs of P.D.A.P.

 These recipes are pictured on the previous page.

ARTICHOKE APPETIZERS

Serves: 6

1 (6.5 oz.) jar marinated artichoke hearts
1 (8.5 oz.) can artichoke hearts, drained
4 green onions, minced
2-3 cloves garlic, minced
1 cup grated Cheddar cheese
1 teaspoon Worcestershire sauce
¼-½ teaspoon Tabasco sauce
¾ cup seasoned bread crumbs
⅓ cup lightly packed minced parsley or cilantro
3 eggs, an additional egg may be used for a more quiche-like texture

Preheat oven to 325°.   8 x 8 baking dish

Grease baking dish. Drain oil from marinated artichoke hearts into a small skillet; add onion and garlic and sauté until tender. Chop artichokes. Mix artichokes, oil, onions, garlic, cheese, eggs, and all seasonings in mixing bowl. Pour into baking dish. Bake for 40-45 minutes. Cut into squares to serve. Cover with foil to reheat.

Can be served warm or cold. Serve with a Hollandaise sauce as a dip if desired.

SPICY COCONUT-CHICKEN BITES

Yield: 50 bites

2 pounds boneless, skinless chicken breasts, cut in 1 inch pieces
2 eggs, beaten
3½ cups sweetened shredded coconut, toasted
1 jar Chef Paul Prudhomme's Poultry Magic spice or Zatarain's Creole Seasoning Mix

2 large cookie sheets

Spray cookie sheets with non-stick spray. Place spices in plastic bag; add chicken pieces, shake to coat. Dip chicken in egg, roll in toasted coconut, and place on cookie sheets. Refrigerate 1 hour. Preheat oven to 400°. Bake 12 minutes, or until golden brown. Serve warm or at room temperature with Dijon mustard.

CRABMEAT IMPERIAL

Serves: 4

1 cup butter or margarine
1 bunch green onions, sliced
1 cup olive oil
1 tablespoon pimientos, chopped

1 pound crabmeat
½ cup sliced fresh mushrooms
⅛ teaspoon salt
⅛ teaspoon pepper
1 tablespoon sherry

Sauté onions in butter for about 5 minutes. Add olive oil, pimiento, crabmeat, mushrooms, salt, and pepper; sauté for about 5 minutes. Add sherry. Serve on toast points.

SAGE STUFFED MUSHROOMS

Serves: 40

1 (10 oz.) package bulk sage sausage
1 tablespoon finely chopped green onion
1 tablespoon chopped fresh parsley

1 clove garlic, minced
¼ teaspoon pepper
1 (8 oz.) package cream cheese, softened, cubed
40 medium mushrooms
⅓ cup butter, melted

Preheat oven to 350°. Shallow baking dish

Brown sausage in a medium skillet until it crumbles; drain. Add onion, parsley, pepper, and garlic. Add cream cheese cubes to the sausage mixture. Remove mushroom stems and discard. Place mushrooms in baking dish. Brush with melted butter. Stuff mushrooms with sausage mixture. Bake for 20 minutes or until bubbly.

 Unpleasant cooking odors can be disguised by simmering a few teaspoons of sugar and cinnamon on the stove.

STUFFED MUSHROOMS

Serves: 18-20

6 cups fresh mushrooms
(3 pints)
¾ cup mayonnaise
Seasoned salt to taste

1 small onion, finely chopped
10 slices cooked bacon,
crumbled
1½ cups grated Cheddar cheese

Preheat oven to 325°.   13 x 9 baking dish

Remove stems from mushrooms and wash in salted water. Drain on paper towels. Mix mayonnaise, seasoned salt, onion, cheese, and bacon. Place mushrooms, cup side up, in baking dish. Fill each mushroom with mixture. Cover with foil. Bake for 15 to 20 minutes. Serve hot.

SAUSAGE CHEESE BISCUITS

Yield: 30 biscuits

1 can buttermilk butterflake
biscuits
2 cups grated Monterey Jack
cheese
1 pound hot sausage, cooked
and drained

2 tablespoons Parmesan cheese
2 eggs
Salt
Pepper

Preheat oven to 325°.   Regular muffin tin

Lightly grease muffin tin. Divide each biscuit into thirds. Place each third into a muffin cup. Mix remaining ingredients together. Place 1 tablespoon of mixture on each biscuit. Bake for 25 minutes. Freezes well. After freezing, thaw, cover with aluminum foil, and warm in slow oven.

 Herbed vinegar set near foods that are frying will help eliminate the odor.

CHERRY TOMATOES WITH CRABMEAT

Serves: 6-8

1 (16 oz.) container of fresh
crabmeat
⅔ cup sour cream
1 (8 oz.) package cream
cheese, softened

1-2 jalapeño peppers, chopped
⅛ teaspoon thyme
2-3 green onions, chopped
2 pints cherry tomatoes
Chopped parsley (optional)

Mix together crabmeat, sour cream, cream cheese, jalapeño peppers, thyme, and green onions. (Mixture can be prepared one night before.) Cut off tomato bottoms and scoop out insides. Let drain. Stuff tomatoes with crabmeat mixture. Sprinkle tops with parsley.

HOT AND SPICY ARTICHOKE DIP

Yield: 2½ cups

1 cup mayonnaise
1 cup grated Parmesan cheese
1 (14 oz.) can artichoke hearts,
drained and chopped
1 (4 oz.) can chopped green
chilies, drained

1 clove garlic, minced
2 tablespoons sliced green
onion
2 tablespoons chopped tomato

Preheat oven to 350°.   9 inch pie plate

Mix all ingredients except onion and tomato. Spoon into pie plate. Bake 20 to 25 minutes or until lightly browned. Sprinkle with onion and tomato. Serve with tortilla chips, crackers, or pita bread wedges.

Microwave: Mix all ingredients except onion and tomato. Spoon into pie plate. Microwave on medium (50%) 6 to 8 minutes or until mixture is warm, stirring every 4 minutes. Stir before serving. Sprinkle with onion and tomatoes. Serve as directed above.

Omit chilies. Stir in 1 (8 oz.) package Crab Delights Salad Shreds.

 Puree ripe papaya with whole grain mustard and a little lime juice. It develops a texture like mayonnaise, so it is great for sandwich spreads.

CURRY DIP

Serves: 4-6

16 ounces cream cheese, soft
½ cup chutney, divided
½ teaspoon dry mustard

2 teaspoons curry powder
½ cup toasted, slivered
almonds (optional)

Whip together cream cheese, 1/4 cup chutney, mustard and curry. Top with remaining chutney and almonds. Serve with a thick, salty chip such as Fritos.

SPICY SHRIMP DIP

Yield: 2 cups

1 (14 oz.) can artichoke
hearts, drained
⅓ to ½ pound shrimp
1 (3 oz.) package cream cheese

½ cup mayonnaise
½ cup picante sauce
¼ cup Parmesan cheese, grated
Thinly sliced green onion tops

Preheat oven to 350°.   9 inch pie plate

Dice artichoke hearts. Add shrimp, cream cheese, mayonnaise, picante sauce and Parmesan. Mix well. Pour into pie plate and bake for 20 minutes. Top with sliced green onions.

SPINACH DIP

Yield: 4 cups

1 (10 oz.) package frozen
chopped spinach
½ cup chopped fresh parsley
1 cup mayonnaise
1 teaspoon white pepper
½ teaspoon dill weed
⅛ teaspoon Tabasco sauce

½ cup chopped onion
2 (3 oz.) packages cream
cheese, softened
1 cup sour cream
1 teaspoon seasoned salt
1 tablespoon freshly squeezed
lemon juice

Thaw and squeeze all moisture from spinach. Place in food processor. Add remaining ingredients and process until well blended. Refrigerate for 6 hours or overnight to blend seasonings.

AVOCADO SHRIMP SPREAD

Serves: 6-8

1 (8 oz.) package cream
cheese, softened
2 tablespoons mayonnaise
Worcestershire sauce to taste
1 clove garlic, minced
1 (6 oz.) tin frozen avocado dip,
thawed

½ cup picante sauce
1 (8 oz.) can baby shrimp,
drained
2 green onions, finely chopped

Cream together cheese, mayonnaise, Worcestershire sauce, and garlic. Spread mixture on a rimmed plate. Cover and refrigerate. Just before serving, frost chilled cheese with avocado dip; then frost with a layer of picante sauce; top with shrimp. Sprinkle with green onions and serve with tortilla chips.

HOT BRIE WITH LAVOSH

Small or large Brie cheese
Butter, softened
Almonds, unpeeled, sliced

Lavosh (Armenian crackers)
or substitute Crackle Snack
or Carr's Water Crackers

Preheat oven to 400°.

Place Brie in dish. Cover generously with softened butter and coat generously with sliced almonds. Bake until almonds are slightly toasted and cheese is piping hot. Serve with Lavosh.

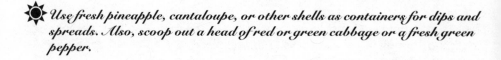

Use fresh pineapple, cantaloupe, or other shells as containers for dips and spreads. Also, scoop out a head of red or green cabbage or a fresh green pepper.

CHUTNEY GLAZED CHEESE PIE

Yield: 2 cups

2 (3 oz.) packages cream
   cheese, softened
4 ounces Cheddar cheese,
   grated
3 tablespoons dry sherry

1-3 teaspoons curry powder
¼ teaspoon salt
½ cup chutney, finely chopped
1 bunch scallions, finely
   chopped

Mix the two cheeses thoroughly. Add sherry, curry powder, and salt; mix. Spread mixture ½ inch thick on serving platter; chill until firm. At serving time spread top with chutney and sprinkle with scallions. Serve with sesame rounds. Looks pretty with green (scallions) and red (pimiento) garnish during the holidays.

FESTIVE CHEESE BALL

2 (8 oz.) packages cream
   cheese, softened
1 (8 oz.) package Cheddar
   cheese, grated
3 green onions with tops,
   chopped
½ green bell pepper, chopped

1 tablespoon lemon juice
1 tablespoon Worcestershire
   sauce
½ teaspoon garlic powder
1 (4 oz.) jar pimientos, chopped
   and drained
Salt to taste

Mix cream cheese, Cheddar cheese, onions, bell pepper, lemon juice, Worcestershire sauce, garlic powder, pimientos, and salt. Form into one large or two small balls. Chill until serving time.

May be rolled in chopped pecans.

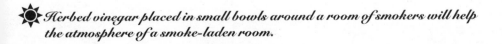

Herbed vinegar placed in small bowls around a room of smokers will help the atmosphere of a smoke-laden room.

GRUYÈRE CHEESE BALL

4 ounces Gruyère cheese,
chilled
¼ clove garlic

4 ounces cream cheese
1½ tablespoons kirsch
Chopped pecans

Use metal blade of food processor. Run machine and drop in cheese. Run 2 minutes, or until finely chopped. Add other ingredients. Form into a ball and roll in pecans. Serve with crackers.

TEXAS CHEESE RING
Yield: 4 cups

1 pound Cheddar cheese, mild
or medium, grated
1 cup mayonnaise
1 cup pecans, chopped

1 small onion, grated
Dash red pepper
Dash black pepper
Strawberry preserves

Combine cheese, mayonnaise, nuts, and onion in a large bowl. Mix well and add peppers. Mold into ring with hands. Chill overnight. Before serving, fill middle of ring with strawberry preserves. Serve cheese on crackers topped with preserves.

SALMON LOG
Yield: 3 cups

1 (1 lb.) can salmon
1 (8 oz.) package cream
cheese, softened
1 tablespoon lemon juice
2 teaspoons grated onion

1 teaspoon horseradish
1 teaspoon liquid smoke
¼ teaspoon salt
½ cup chopped pecans

Drain and flake salmon. Add cheese, lemon, onion, horseradish, liquid smoke, and salt. Mix well. Chill overnight. Shape into log. Chill 2-3 hours longer. Sprinkle pecans on waxed paper and roll log in pecans. Chill until served.

 Silver will gleam after rubbing with baking soda on a soft damp cloth.

SOUTH TEXAS CHEESECAKE

Serves: 20

Crust:

1 cup tortilla chips, finely
crushed

3 tablespoons butter, melted

Preheat oven to 325°.

9 inch springform pan

Stir chips and butter in small bowl; press onto bottom of springform pan. Bake 15 minutes.

Filling:

2 (8 oz.) packages cream
cheese, softened
2 eggs
1 (8 oz.) package Colby/
Monterey Jack cheese,
shredded

1 (4 oz.) can chopped green
chilies, drained

Beat cream cheese and eggs in large mixing bowl at medium speed with electric mixer until well blended. Mix in cheese and chilies; pour over crust. Bake 30 minutes.

Toppings:

1 cup sour cream
1 cup chopped yellow, orange,
or red bell pepper
½ cup sliced green onion

⅓ cup chopped tomatoes
¼ cup sliced pitted ripe olives or
red pimientos

Spread sour cream over cheesecake. Loosen cake from rim of pan; cool before removing rim. Refrigerate several hours or overnight. Top with remaining ingredients. Serve with tortilla chips or crackers.

 Cover a block of cream cheese with jalapeño jelly and serve with crackers. For a different touch, press canned crabmeat into a block of cream cheese and pour cocktail sauce over it.

HOLIDAY APPETIZER PIE

Yield: 20 servings

1 (8 oz.) package cream
cheese, softened
2 tablespoons milk
2½ ounces dried beef, finely
chopped
⅓ cup chopped onions
3 tablespoons chopped green
bell pepper

⅛ teaspoon black pepper
⅛ teaspoon garlic powder
1 cup sour cream
¼ cup chopped pecans, sautéed
in butter

Preheat oven to 350°.   8 inch glass pie pan

In a small bowl mix well cream cheese, milk, and dried beef. Stir in onion, bell pepper, black pepper, and garlic powder. Add sour cream. Spoon into pie pan. Sprinkle with chopped pecans. Heat in oven for 15 minutes (or microwave on high 1-2 minutes). Serve with crackers.

HORS D'OEUVRE PIE

Serves: 75-100

6 (8 oz.) packages cream
cheese, softened
1 (8 oz.) package bleu cheese,
crumbled
1 (8 oz.) package Cheddar
cheese, grated
½ cup minced chives

1 green onion, minced
Red caviar (or drained
pimientos)
Black caviar
Chopped green olives
Chopped hard-boiled eggs

10 inch springform pan

Place the following three layers separately in springform pan: Blend 16 ounces cream cheese with bleu cheese to form bottom layer; blend 16 ounces cream cheese with grated Cheddar cheese to form middle layer; and blend 16 ounces cream cheese with minced chives and minced green onion to form top layer. Each layer can be blended separately in a food processor. You may place layers on top of each other without waiting for them to set. A few hours before serving, score top of pie into 8 wedges. After removing the sides of the springform pan, decorate top of pie with the following ingredients (place each ingredient on two opposite sections of pie): red caviar (or drained pimientos), black caviar, chopped hard-boiled eggs, and green olives. Serve with crackers.

PARTY PATÉ

16 ounces liverwurst
4 ounces cream cheese
4 teaspoons sherry

⅛ teaspoon curry powder
¼ cup raisins (scald with boiling water and drain)

Mix all ingredients. Form a mound and serve with thin slices of French bread.

WATERMELON ICEE

Serves: 8

½ large watermelon (may substitute other fruit)
2 tablespoons honey

Crushed ice or ice cubes
Fresh mint, for garnish

Cut watermelon in chunks, removing black seeds. Fill blender with watermelon; add honey. Blend briefly to mix. Add ice cubes and blend to desired consistency. Serve in a stemmed glass with mint garnish.

Good with cantaloupe, too!

Very healthy—no sugar.

SOUTH TEXAS BLOODY MARY MIX

Serves: 15-20

2 (40 oz.) cans tomato juice
2 (10 oz.) cans beef broth
5 ounces lemon juice
5 ounces Worcestershire sauce
1 quart orange juice

2 dashes Tabasco sauce
3 ounces salt
Vodka
Ice

Mix all ingredients except vodka and ice. Add desired amount of vodka and serve over ice.

BRANDY ALEXANDER

Serves: 8-10

½ gallon softened coffee ice cream
¼ cup brandy
¼ cup Creme de Cacao liqueur
½ pint whipping cream
1 teaspoon vanilla extract

1 teaspoon cream of tartar
¼ cup sugar
1 small chocolate bar
1 box Pepperidge Farm champagne cookies

Chill a 2-3 quart bowl and egg beaters that are grease free.

Mix brandy and Creme de Cacao liqueur into ice cream. Pour into stemmed glasses and freeze. Using chilled bowl and beaters, whip the whipping cream, vanilla, and cream of tartar. Gradually add the sugar and continue whipping until the cream forms stiff peaks. Set aside. Make chocolate shavings using a potato peeler or knife. Before serving, top with whipped cream and garnish with the chocolate shavings. Serve with champagne cookies and a straw.

CHAMPAGNE PUNCH

Serves: 24

2 cups sugar
2 cups water
Juice of 6 lemons
2 cups apricot nectar
1 (6 oz.) can frozen orange juice concentrate

2 (12 oz.) cans frozen apple juice concentrate
2 cups pineapple juice
2 (12 oz.) bottles ginger ale, chilled
2 fifths champagne, chilled

Punch bowl

Boil sugar and water for one minute and cool; add juices and freeze in a Bundt pan or a similar mold. Thaw 1-1½ hours before serving. Add cold ginger ale and champagne. Serve over block of ice in punch bowl or substitute frozen ring mold of punch juices. For non-alcoholic punch, use all ginger ale.

 To remove caffeine stains or cigarette burns from fine china, rub with a damp cloth dipped in baking soda.

FROZEN DAIQUIRI

Yield: 5 cups

6 ounces frozen limeade, thawed

12 ounces rum
18 ounces water

Combine limeade, rum, and water, place in a covered plastic container; freeze. Remove from freezer about ½ hour before serving. Stir and spoon into champagne glasses.

Strawberry variation: Substitute frozen lime juice for limeade and add 10 ounces frozen strawberries. Mix in blender and freeze.

RUM EGGNOG

Serves: 10

6 eggs, separated
¼ cup sugar
1 cup light rum

1 quart milk
1 pint ice cream
Nutmeg

Beat egg yolks. Add sugar to yolks and blend thoroughly; add rum and milk, mixing well. Fold in stiffly beaten egg whites; add ice cream. Serve in punch cups with a dash of nutmeg.

MILK PUNCH

Serves: 16

1 cup sugar
1 cup water
½ gallon whole milk

1 pint bourbon
½ teaspoon vanilla extract
1 block ice

Make a simple syrup by combining sugar and water and heating in a double boiler until sugar dissolves. Cool. Cover bottom of a 1 cup measuring cup with simple syrup. Reserve extra syrup for use in doubling recipe. Add syrup to milk, bourbon, and vanilla. Serve in punch bowl. Add block of ice just prior to serving. May be made a day ahead and stored in refrigerator.

MIMOSA PUNCH

Serves: 24

5 large oranges
15 maraschino cherries
Lemon leaves
⅓ cup sugar
3 (2½ inch) cinnamon sticks
3 cups orange juice
¼ cup orange flavored liqueur

¼ cup brandy
1 (1 liter) bottle club soda, chilled
1 (750 ml.) bottle champagne or sparkling white wine, chilled

6 quart punch bowl   6 cup ring mold

One day in advance prepare ice ring. With small knife, cut continuous 1 inch wide strip of peel from each orange; reserve peels. Squeeze juice from oranges; add water to orange juice to measure 5 cups liquid. Pour 3½ cups orange juice mixture into 6 cup ring mold; freeze about 3 hours. Refrigerate remaining juice mixture.

Meanwhile, carefully trim white membrane from orange peel strips. Form each orange-peel strip into a rose by rolling it tightly, skin-side out; cover and refrigerate.

When juice ring is frozen, arrange orange-peel roses, maraschino cherries, and lemon leaves on frozen mixture for garnish; pour in half of remaining mixture; freeze until firm to set garnishes in place, about 1 hour. Add remaining juice mixture (make sure orange peel-roses, cherries, and lemon leaves extend above juice). Freeze.

In 1 quart saucepan over high heat, heat sugar, cinnamon, and ½ cup water to boiling. Reduce heat to low; cover and simmer 15 minutes. Refrigerate until ready to complete punch.

Just before serving, in a punch bowl large enough to hold ice ring, mix orange juice, liqueur, brandy, and cinnamon mixture. Stir in soda and champagne.

To serve, place punch bowl on tray. Unmold ice ring; add to punch. Decorate tray with leaves and ribbon, depending on occasion. Doubles easily.

A delicately spiced champagne. Spectacular. Perfect for a holiday toast or bridal party.

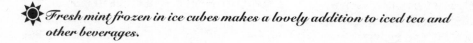

Fresh mint frozen in ice cubes makes a lovely addition to iced tea and other beverages.

PINEAPPLE SANGRÍA

Yield: 2½ quarts

1 (20 oz.) can pineapple
  chunks, reserve juice
  Wooden skewers
1 apple, unpeeled, cored, and
  cut into chunks
1 lemon, sliced and quartered

1 orange, sliced and quartered
1 (750 ml.) bottle Burgundy
  wine
1 (46 oz.) can pineapple juice
¼ cup brandy
¼ cup sugar

Skewer drained pineapple chunks, apple chunks, lemon, and orange slices onto wooden picks in an assorted fashion. Combine reserved pineapple juice, Burgundy, canned pineapple juice, brandy, and sugar in a container. Stir well. Add skewered fruit kabobs. Refrigerate at least 1 hour. Serve the drinks over ice with a fruit kabob in each glass.

SANGRÍA

Serves: 10

1 quart dry red wine
½ cup sugar
2 lemons, juiced

2 oranges, juiced
2 jiggers brandy
1 (7 oz.) bottle sparkling water

Mix wine, sugar, lemon juice, orange juice, and brandy. Chill. Add sparkling water and serve over ice cubes.

SUNSHINE SPECIAL

Serves: 6

1 cup gin or vodka
2 cups fresh orange juice
1 cup sugar
3 eggs

3 tablespoons powdered coffee
  creamer
Juice of 3 large lemons,
  approximately 9 tablespoons

Mix all ingredients in a pitcher with 6-7 ice cubes. Blend about half of this mixture at a time in a blender. Serve very cold.

Variation: ¼ cup of dairy cream may be substituted for the powdered coffee creamer.

SPICED CIDER

Serves: 8

2 quarts apple cider
1 teaspoon whole cloves
1 teaspoon allspice

3 sticks cinnamon
½ unpeeled lemon, thinly sliced
¼-½ cup sugar

Simmer all ingredients together. Serve hot. Remove spices and refrigerate any leftovers. May be reheated. Great on cold days!

COFFEE PUNCH

Yield: 8 quarts

½ cup instant coffee
2 cups sugar
3 quarts hot water
2 quarts milk

1 (5.5 oz.) can of chocolate syrup
1 gallon vanilla ice cream, softened

Combine coffee, sugar, and water; cool. Add milk, syrup, and ice cream. Freeze. Remove 2 to 2½ hours before serving. Place mixture in punch bowl to serve.

FRUITY PUNCH

Serves: 50

2 (3 oz.) packages cherry gelatin
13 cups water
6 cups water
6 cups sugar

3 (46 oz.) cans unsweetened pineapple juice
1 large bottle lemon juice
1 ounce almond extract
3 quarts ginger ale

Boil 13 cups water; add gelatin, mix well, and cool. Boil 6 cups water; add 6 cups sugar, mix well, and cool. Add pineapple juice, lemon juice, and almond extract to the mixtures. Freeze until firm. To serve: partially thaw and place in punch bowl; break into chunks; add ginger ale, and stir until slushy.

GOLDEN PUNCH

Serves: 10-20

2 (20 oz.) cans crushed
pineapple in juice
2 (6 oz.) cans frozen
lemonade (undiluted)

¼ cup sugar
1 (28 oz.) bottle club soda,
chilled

In blender, blend pineapple one can at a time. Add lemonade and sugar; blend well. Place mixture in punch bowl; add club soda and stir.

OPENING RECEPTION PUNCH

Serves: 100

5 pounds sugar, 10 cups
4 cups water
2 tablespoons citric acid
1 large bottle lemon juice

2 ounces vanilla extract
2 ounces almond extract
Ginger ale

Heat the sugar, water, citric acid, and lemon juice until dissolved. Add the vanilla and almond extract. This mixture is a concentrated syrup. It can be refrigerated or frozen until ready to use. When ready to use, mix 1½ cups concentrate with 3 bottles of ginger ale. Wonderful as a mock champagne punch for a wedding.

ORANGE FIZZ

Serves: 4

1 (6 oz.) can frozen orange
juice
1 cup milk

1 cup water
¼ cup sugar
1 teaspoon vanilla extract

Thaw can of orange juice. Place all the ingredients in blender; add ice cubes to fill. Blend to desired consistency.

FROZEN STRAWBERRY COOLER

Serves: 4-6

1 (6 oz.) can frozen limeade

1 small box of frozen strawberries

Blender

Fill blender with ice. Add limeade and strawberries. Blend until smooth. Add water if needed.

SUNSHINE PUNCH

Serves: 15

1 (6 oz.) can frozen orange juice concentrate
1 (6 oz.) can frozen lemonade concentrate

1 (6 oz.) can frozen limeade concentrate
4 cups water
1 liter ginger ale

Punch bowl

Combine first four ingredients. Mix well. Stir in ginger ale just before serving.

TROPICAL ICE PUNCH

Serves: 12

2 cups mashed bananas
1 (20 oz.) can crushed pineapple, undrained
1 (17 oz.) carton frozen strawberries in syrup

2 cups orange juice
1 tablespoon lemon juice
1 cup sugar
1 (33 oz.) bottle ginger ale

Combine bananas, pineapple, frozen strawberries, orange juice, lemon juice, and sugar. Stir well and freeze until firm. To serve, partially thaw fruit mixture. Place in punch bowl and break into chunks. Add ginger ale and stir until slushy. Fruit mixture can be frozen up to one month ahead, if well covered.

Blender clean up: Fill blender part way with hot water; add a drop of detergent, cover, turn on for a few seconds. Rinse and drain dry.

Soups
& Salads

Refer to Mexican Cuisine section for additional recipes in this category.

McAllen International Museum (MIM)

With hard work and generous donations the Junior League of McAllen, Inc. provided the driving force to establish MIM. Since then, we have given our wholehearted support through donations exceeding $100,000. Part of the proceeds of our first cookbook, *La Piñata*, provided the museum's landscaping. Our museum volunteer work, which has included docenting, puppet shows, the Christmas Tree Forest, and art shows, has helped bring the arts to thousands who have visited the museum and raised extra funds for MIM.

 These recipes are pictured on the previous page.

CHILLED AVOCADO SOUP

Serves: 12-14

4 medium ripe avocados, peeled and pitted
1 clove garlic, minced
4 green onions, chopped
5 tablespoons chopped fresh cilantro
1 tablespoon sliced pickled jalapeños

½ teaspoon Tabasco sauce
3 cups sour cream
1 cup buttermilk
8 cups chilled chicken broth
Salt to taste

In a blender or food processor, combine avocados, garlic, green onions, cilantro, jalapeños with juice, and Tabasco. Process until smooth. Add sour cream; process again. Stir in buttermilk and chilled chicken broth. Taste for salt. Cover; refrigerate until very cold.

May be garnished with a dollop of sour cream and chopped green onions (green part only).

COLD CHERRY SOUP

Serves: 4

2 cups water
½ cup sugar
½ teaspoon salt
1 cinnamon stick
Juice of ½ lemon
1 (16 oz.) can dark sweet cherries with juice

1-1½ tablespoons flour
¾ cup sour cream
¾ cup red wine
2-4 tablespoons Chamboid liqueur, optional

In a large pan combine water, juice from cherries, sugar, salt, lemon juice, and cinnamon stick. Boil for 3 to 4 minutes; taste liquid for flavoring. If not well-flavored, let boil and reduce for another few minutes. Add the cherries to the boiling mixture; simmer 4-5 minutes. Mix the flour, sour cream, and wine until smooth. Slowly stir in a ladle of hot cherry juice. Pour mixture into the soup; boil until it thickens. Add Chamboid, if desired. Cool. Serve chilled.

Ice cubes will eliminate the fat from soup and stew. Just drop a few into the pot and stir; the fat will cling to the cubes; discard the cubes as they melt. Or, wrap ice cubes in paper towel or cheesecloth and skim over the top.

GAZPACHO

Serves: 8-10

4 tomatoes
3 cucumbers
2 stalks celery
1 small onion
1 bell pepper
3 cups tomato or vegetable
   juice
3 tablespoons olive oil
2 tablespoons lemon juice
1 teaspoon sugar

1 teaspoon Worcestershire
   sauce
¼ teaspoon Tabasco sauce
3-4 tablespoons red wine vinegar
   Salt to taste
4 teaspoons chopped parsley
2 small cloves garlic, minced
4 teaspoons chopped cilantro
½ teaspoon freshly ground
   pepper

Chop all vegetables. (If using a food processor, chop each separately; be sure not to over-chop. They should be a bit chunky.) Combine with remaining ingredients. Refrigerate until cold. Serve in chilled bowls, mugs, or any fun serving piece. Top with avocado and sour cream; or with chopped herbs such as basil, cilantro, parsley, or chives.

CHICKEN NOODLE SOUP

Serves: 6-8

2½ pounds chicken breasts
1 quart water
2 cups carrots,
   cut into ½ inch slices
2 cups celery,
   cut into ½ inch slices

1 teaspoon sugar
¼ teaspoon pepper
3 chicken bouillon cubes
1½ cup uncooked thin noodles

Heat all ingredients except noodles to boiling; reduce heat. Cover and simmer until chicken is done, about 45 minutes. Skim fat if necessary. Cook noodles per package directions. Remove chicken from bones and skin. Cut chicken into one-inch pieces. Add chicken and noodles to broth; heat until hot, about 5 minutes.

RED PEPPER VICHYSSOISE

Serves: 6-8

Soup base:

2 tablespoons butter
2 cups finely sliced leeks
3 cups finely sliced potatoes
4½ cups chicken stock

1 cup heavy cream
Salt and white pepper to taste
Chopped chives

Sauté leeks in hot butter until soft. Add potatoes and stock; simmer partially covered until potatoes are tender, about 30 minutes. Puree soup; add cream. Season to taste.

Red Pepper Puree:

3-4 red bell peppers

Cut peppers into thirds. Boil 6 minutes. Remove skins, seeds, and membranes. Puree.

Crème fraîche:

2 tablespoons buttermilk

1 cup heavy cream

Add buttermilk to cream. Cover; set overnight at room temperature. It will thicken nicely and have the flavor of sour cream.
Stir pepper puree into soup base. Garnish with chopped chives and a dollop of crème fraîche, if desired.
You may substitute commercial sour cream for crème fraîche.

 Storing leftover chicken or beef stock? Freeze it in ice cube trays, remove cubes and place in sealable bags in the freezer.

SOUTHWESTERN BEAN SOUP

Serves: 6-8

8 slices bacon, diced
1½ cup chopped onion
1½ cup chopped celery
2 cloves garlic, minced
2 (16 oz.) cans refried beans
2 (13¾ oz.) cans chicken broth

2 (4 oz.) cans chopped green chiles, undrained
½ teaspoon chili powder
½ teaspoon pepper
Shredded Cheddar cheese
Crushed tortilla chips

Cook bacon in large saucepan until crisp; remove bacon and reserve drippings in saucepan. Set aside bacon. Saute onion, celery and garlic in drippings over medium heat, stirring constantly, until tender. Add next 5 ingredients and stir well. Bring to boil, reduce heat and simmer uncovered 10 minutes. Sprinkle each serving with bacon, cheese and chips.

SWEET AND SOUR CABBAGE SOUP

Serves: 8

1 package beef soup bones
1 package beef soup marrow bones
4 quarts water
1 onion, sliced into rings
1 head cabbage, shredded
1 (14½ oz.) can stewed tomatoes

1 tablespoon lemon juice
¾ cup sugar
2 tablespoons salt
1 (8 oz.) can tomato sauce
Sour cream (optional)

Cover soup bones with water in large pot. Bring to a boil; skim foam off top. Add all other ingredients. Simmer about 3 hours. Taste periodically to adjust sweet and sour flavors.

Delicious served with black bread and a dollop of sour cream atop each serving. Freezes.

BOUILLABAISSE

Serves: 4

1½ cups finely chopped onion
1 cup thinly sliced leek, white part only
1 teaspoon finely chopped garlic
2 cups fish or chicken broth, defatted
1½ cups peeled, seeded, diced tomatoes
2 tablespoons finely chopped parsley
½ cup finely chopped celery
1 bay leaf
½ teaspoon crushed dried thyme

¼ teaspoon crushed dried fennel
¼ teaspoon dried saffron, dissolved in a little stock
⅛ teaspoon freshly ground black pepper
1 cup dry white wine
½ pound shellfish, shelled and cleaned
½ pound firm boneless white fish, cut into strips
Fresh thyme or fennel for garnish (optional)

Combine onions, leeks, and garlic. Cook, covered, over very low heat until soft, about 10 minutes. Add a little stock if necessary to prevent scorching. Add remaining ingredients, except shellfish and fish. Mix well; bring to a boil. Reduce heat and simmer, covered, for 10 minutes. Add fish and shellfish. Cook until fish turns from translucent to opaque, for 2-5 minutes. Garnish with thyme or fennel.

Serve with crusty French bread.

This is low calorie (240) and low fat.

CHEESE SOUP

Serves: 6-8

4 tablespoons margarine
3 scallions, chopped
3 ribs of celery, chopped
2 carrots, grated
1 (10¾ oz.) can low-salt chicken broth and
1 can beer (or 2 cans broth)

3 cans low-salt potato soup
8 ounces Cheddar cheese, grated
Dash of white pepper, Tabasco sauce, and parsley
Sour cream (optional)

Over low heat, sauté onions, celery, and carrots in margarine. Add chicken broth, cover, and simmer for 20 minutes. Add potato soup, cheese, parsley, Tabasco, and pepper. If using beer, add after potato soup and cheese according to thickness of soup preferred. May also add 8 ounces of sour cream, if desired. Simmer 15 minutes.

CHICKEN VEGETABLE GUMBO— SOUTH TEXAS STYLE

Serves: 12

2 whole chickens
½ head cabbage, shredded
3 large onions, chopped
1 (16 oz.) can peeled tomatoes or fresh
1 stalk celery, chopped
1 package carrots, sliced
1 bunch cilantro, chopped
1 (16 oz.) package frozen corn

1 (16 oz.) package frozen cut green beans
1 cup raw rice
4 tablespoons black pepper
4 tablespoons garlic powder or cloves of garlic
Texjoy seasoning to taste
2 tablespoons oregano

Boil chicken in 1 gallon of water. Remove chicken. Add all ingredients, except rice; cook 30 minutes to 1 hour. Bone chicken. Add chicken and rice. Bring to a boil. Simmer until rice is cooked.

CRAB AND BROCCOLI SOUP

Serves: 4-6

½ cup chopped onion
3 tablespoons butter, melted
2 tablespoons flour
2 cups milk
2 cups light cream
2 chicken bouillon cubes
½ teaspoon salt

¼ teaspoon thyme
⅛ teaspoon pepper
⅛ teaspoon red pepper
1 (10 oz.) package frozen chopped broccoli, cooked and drained
8 ounces crabmeat

3 quart saucepan

Sauté chopped onion in butter. Blend in flour. Add milk and cream; heat thoroughly, stirring constantly. Dissolve bouillon cubes in hot soup. Add seasonings, broccoli, and crabmeat. Heat gently. Serve with French bread and white wine.

 To get the best out of the delicate flavors of herbs, put them into the preparation only in the last few minutes of cooking. Prolonged cooking accounts for an unaromatic or bitter taste.

LIME SOUP

Serves: 4

4 cups chicken broth
2 chicken breast halves,
   skinned
⅓ cup alphabet pasta
¼ cup chopped red onion

¼ cup fresh lime juice,
   (1-2 limes)
⅓ cup chopped parsley
1 small ripe avocado, peeled,
   pitted, and sliced

In a saucepan, bring broth to a boil. Add chicken; reduce heat to simmer; cover and cook 20 minutes. Remove chicken from broth; cool slightly. Bone chicken and shred. Set aside. Reheat broth to boiling; add pasta. Reduce heat; cook for 10 minutes. Add onion, lime juice, chicken, and parsley. Add avocado. Cover and simmer for 5 minutes.

MULLIGATAWNY SOUP

Serves: 4-6

1 medium onion, sliced
1 medium carrot, peeled and
   diced
1 stalk celery, diced
1 cup cooked chicken bits
1 medium tart apple, cored,
   pared, and sliced ¼ inch thick
¼ cup margarine
⅓ cup flour
1 teaspoon curry powder

Dash of mace
2 whole cloves
1 sprig parsley
1 cup canned tomatoes
2 chicken bouillon cubes
1 teaspoon salt
2 cups boiling water
Dash of pepper
3 cups milk or half-and-half

Deep kettle or Dutch oven

Sauté onions, carrots, celery, chicken, and apple in margarine in deep kettle or Dutch oven, stirring frequently, until onion is tender. Stir in remaining ingredients, except milk or half-and-half. Simmer covered, ½ hour or until tender. Just before serving add milk or half-and-half; heat thoroughly, but do not boil.

May be made with skim milk and diet margarine.

 Drop a lettuce leaf into a pot of homemade soup to absorb excess grease from the top.

MINESTRONE SOUP

Serves: 10

1 cup dried navy beans
2 (13¾ oz.) cans clear chicken
broth
2 teaspoons salt
1 small head cabbage, shredded
4 carrots, peeled and sliced
2 medium potatoes, peeled and
diced
1 (28 oz.) can Italian style
tomatoes, chopped

2 medium onions, chopped
1 rib of celery, chopped
1 large fresh tomato, peeled
and chopped
1 clove garlic, minced
¼ cup olive oil
1 cup broken thin spaghetti
Chopped parsley
¼ teaspoon pepper

Soak beans overnight in water to cover. Measure chicken broth and add water to measure 3 quarts. Add salt and cook beans in broth mixture until almost tender. Add cabbage, carrots, potatoes, and tomatoes. Sauté onions, celery, fresh tomatoes, and garlic in olive oil and add to soup. Add spaghetti, parsley, and pepper. Simmer until spaghetti is tender. Serve with spoonful of Pesto sauce per bowl. The Pesto sauce can be prepared ahead.

Pesto Sauce:

¼ cup butter, softened
¼ cup Parmesan cheese
½ cup chopped parsley
1 clove garlic, minced
1 teaspoon crushed basil leaves

½ teaspoon crushed marjoram
leaves
¼ cup olive oil
¼ cup chopped pine nuts or
walnuts

Mix all ingredients together.

 Chop fresh parsley and put it into a sectional ice cube tray, then fill with water and freeze. Just drop the cubes into soups, stews, etc., as needed.

MUSHROOM SOUP

Serves: 4-6

4 tablespoons unsalted butter
1 medium onion, finely chopped
1 pound mushrooms, coarsely
  chopped
3 tablespoons all-purpose flour

4 cups beef bouillon or beef
  stock
Dash of white pepper
Dash of nutmeg
1½ cups whipping cream

Melt butter in a large saucepan. Add chopped onion and sauté until clear. Add mushrooms; cook until tender. Add flour, stirring constantly. Stir in bouillon. Bring to a boil; stir in pepper and nutmeg. Remove from heat; stir in cream. Serve immediately. May be prepared in advance by omitting the cream. Refrigerate. To serve, just reheat and add cream.

FRENCH ONION SOUP

Serves: 2

1 tablespoon butter
1½ teaspoons olive oil
½ pound onions, thinly sliced
¼ pound leeks (white part only),
  sliced
1 clove garlic, chopped
¼ teaspoon sugar
1½ tablespoons flour

4 cups canned beef broth
¼ cup dry vermouth
Dash of Tabasco sauce
Ground pepper
Salt
Toasted French bread slices
Thinly sliced baby Swiss
cheese

Melt butter with oil in a large pot over low heat. Add onions, leeks, and garlic. Cook until onions soften, stirring occasionally, about 15 minutes. Increase heat to medium-low. Add sugar; cook until onions are a deep golden brown, stirring frequently, about 30 minutes. Add flour; stir for 3 minutes. Add stock, vermouth, Tabasco, and pepper. Simmer 40 minutes. Season with salt. Can be prepared 1 day ahead to this point, but re-warm before continuing. Divide soup between 2 soup crocks. Top each with toasted bread and several slices of cheese. Broil until cheese melts and top is golden brown.

CREAM OF POTATO SOUP

Serves: 10

5 cups chopped potatoes
½ cup chopped carrots
½ cup chopped celery
½ cup chopped onion
4-5 cups chicken broth
3 teaspoons Beau Monde seasoning

½ teaspoon pepper
Salt to taste
1 cup instant potatoes
½ pint whipping cream
Milk

In a stockpot, combine potatoes, carrots, celery, onion, and cover with chicken broth. Bring vegetables to a boil, reduce heat, and simmer until tender. Add Beau Monde, pepper, and salt. Simmer for 5-10 minutes. Remove from heat. Add instant potatoes; stir. Add whipping cream; stir. Thin soup to desired thickness with milk.

BRANDIED PUMPKIN SOUP

Serves: 6-8

¼ cup butter
½ cup finely chopped onion
¼ teaspoon ginger
¼ teaspoon nutmeg
3½ cups chicken broth
2 (16 oz.) cans pumpkin
1 cup whipping cream

1 cup half-and-half
Salt and pepper
2 tablespoons brandy, or to taste
Sour cream
Pepitas, shelled (dried pumpkin seeds)

5 quart stock pot

Melt butter in the pot. Add chopped onion and cook, stirring occasionally until transparent. Blend in ginger, nutmeg, and chicken broth. Bring to a boil. Blend in pumpkin, whipping cream, and half-and-half. Reduce heat; cook until soup is heated, stirring occasionally. Season to taste with salt and pepper; add brandy, if desired. Garnish with sour cream. Top with Pepitas.

 Instead of using cream to thicken soups, use the main ingredient as a thickener. Puree some vegetables for a healthier soup.

FRESH SQUASH SOUP

Serves: 16

¼ pound lean bacon, diced
1 large onion, chopped
2 cloves garlic, chopped
8 large yellow squash, sliced
2 large zucchini, sliced
8 cups chicken broth
5 teaspoons cumin

1½ teaspoons oregano
4 cups evaporated milk
Salt and pepper to taste
Grated Cheddar cheese
Chopped green onion
Pimiento strips

5 quart stock pot

In a large stock pot, brown diced bacon. Discard all but 2 tablespoons of bacon grease. Add onion; sauté until transparent. Add garlic, yellow squash, zucchini, chicken broth, and spices. Bring to a boil. Reduce heat, simmer covered, for 10 minutes. Add evaporated milk; heat. Using a potato masher, coarsely mash squash. Garnish with grated Cheddar, chopped green onion, and pimiento strips.

May add sliced green chilies, corn, and chicken.

SQUASH AND GREEN CHILI SOUP

Serves: 8

1 (14½ oz.) can chicken broth
4 chicken bouillon cubes
2½ pounds zucchini and yellow squash, sliced
½ medium onion, finely chopped

1 (4 oz.) can green chilies, chopped
1½ tablespoons butter
⅛ teaspoon pepper
1 (8 oz.) carton sour cream

Simmer squash in chicken broth and bouillon cubes in covered pot for 20 minutes, or until squash is soft. Add onion and green chilies and cook 5 minutes more. Put mixture through food processor, but do not puree completely. The soup should have body and little flecks of green. Add butter, pepper, and sour cream.

Good either hot, warm, or cold. This may be thinned with water or milk, if desired.

SEAFOOD GUMBO

Serves: 8-10

2 pounds shrimp, shelled and cleaned
2 bunches green onions, chopped
1 large onion, chopped
1 bell pepper, chopped
2 cups chopped celery
1 cup cut okra

1 (10 oz.) can diced tomatoes with green chilies
1 (48 oz.) can tomato juice
Tabasco to taste
1 pound crabmeat
4 tablespoons roux
Cooked rice

Large stock pot

Wash shrimp shells well; cook in large pot of water. Boil for an hour or so adding water as needed to make a good shrimp broth*. Use this in place of water.
Sauté vegetables in 2-4 tablespoons butter, or shrimp broth until tender; add tomatoes and juice. Simmer about 20 minutes and add one can (48 oz.) water*; simmer another 20 minutes and spoon roux mixture into gumbo soup stirring constantly to keep from lumping. Add shrimp, crab, (or any seafood you prefer) and cook until shrimp are pink.

Roux:
Melt 3 tablespoons butter in small saucepan. Stir in 4 heaping tablespoons flour. Stir constantly over medium high heat until it is a caramel brown, almost burning. Add one cup of gumbo to roux, stirring well to blend, add to gumbo.

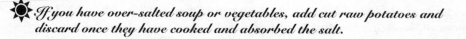

 If you have over-salted soup or vegetables, add cut raw potatoes and discard once they have cooked and absorbed the salt.

VEGETABLE SOUP

Serves: 6-8

¾ cup barley
¾ cup green split peas
½ cup dried white beans
2 pounds beef chuck
1 package marrow bones
3 carrots, diced
2 cups diced celery
1 bell pepper, diced
1 (28 oz.) can peeled tomatoes
1 onion, whole

2 turnips, diced
1 large white potato, peeled and diced
1 parsnip, whole
2 sweet potatoes, diced
1 yellow squash, diced
1 green squash, diced
¼ cup alphabet pasta
1½ cups frozen vegetables
Parsley to taste

Large soup pot

Rinse barley, split peas, and white beans in a colander. Drain. Soak beans, peas, and barley overnight in water. The next day put beef, bones, and beans in a large soup pot with water; bring to a boil. Simmer on medium heat for 1 hour. Add carrots, parsnip, turnips, sweet potatoes, white potato, squash, celery, green pepper, canned tomatoes, and onion. Cook for 45 minutes; add parsley, frozen vegetables, and alphabet pasta. Cook additional 15 minutes.

DIJON SALAD DRESSING

Yield: 2½ cups

½ cup vinegar (white or wine)
¼ cup Dijon mustard
1 teaspoon salt
1 tablespoon dried onion flakes
1 tablespoon dried parsley

1 teaspoon garlic powder
1½ cups vegetable oil or (¾ cup vegetable oil and ¾ cup olive oil)

Place all ingredients except oil in a jar. Shake well. Add oil. Shake again. Taste. Add additional mustard to taste. Chill and serve.

GARLIC SALAD DRESSING

Serves: 4

5 cloves garlic, freshly crushed
1 tablespoon Worcestershire
  sauce
1 tablespoon Gulden's mustard

¼ cup red wine vinegar
⅔ cup sunflower oil
*1 tablespoon honey
  *Optional for sweet dressing

Mix all ingredients in a salad dressing shaker top bottle. Shake well and serve chilled.

CREAMY HONEY LIME DRESSING

Yield: 1¾ cups

1 cup sour cream
1½ cups cottage cheese

¼ cup fresh lime juice
2 tablespoons honey

In a blender, whirl sour cream, cottage cheese, lime juice, and honey until well blended. Chill; serve over fresh fruit.

HONEY MUSTARD SALAD DRESSING

1½ cups safflower oil
1 cup white vinegar
¾ cup honey

8 tablespoons Dijon mustard or
  ¾ of an (8 oz.) jar
1 teaspoon poppy seeds

Whip or shake the oil and vinegar together in a jar. Mix in the remaining ingredients.

A lemon heated in hot water for 2 minutes or microwaved for 30 seconds will yield 2 tablespoons more juice than an unheated one. This applies to all citrus fruits.

MADRAS DRESSING

Serves: 20

1 quart mayonnaise
1 small onion, finely chopped
1 teaspoon dry mustard
2 tablespoons mustard
½ teaspoon white pepper

½ teaspoon MSG (optional)
Dash of salt
Curry powder to taste
Romano or Parmesan cheese

Mix mayonnaise, onion, dry mustard, mustard, pepper, MSG, salt, and curry powder together. Add a dash of Romano or Parmesan cheese if desired. Dressing keeps in refrigerator for quite awhile.

CAESAR SALAD

Serves: 6

1 clove garlic
½ cup virgin olive oil, divided
1 cup croutons
1 large head romaine lettuce
¼-½ teaspoon salt
Freshly ground pepper

1 one minute egg
Juice of one lemon
3-4 anchovy fillets, chopped
¼ cup freshly grated Parmesan cheese

Crush garlic in small bowl; pour oil over and let stand several hours. Brown croutons (made from stale French bread) in 3 tablespoons garlic oil, stirring often, or toast in slow oven. Tear romaine into large salad bowl; sprinkle with salt and ground pepper. Pour remaining garlic oil over, and toss until every leaf is glossy. Break the one minute egg into salad; squeeze lemon juice over and toss thoroughly. Add chopped anchovies and grated cheese, toss again. Add croutons; toss gently and serve immediately.

 Brush oil on a grater before shredding cheese for an easy clean-up.

ELEGANT SALAD

Serves: 8

Salad:

2 heads Bibb lettuce
2 bunches scallions, sliced
½ cup chopped, toasted almonds
4 avocados, diced

2 (4 oz.) cans mandarin oranges, drained
4 tablespoons capers
1 (12 oz.) jar artichoke hearts, drained and sliced

Wash and dry lettuce; tear into medium sized pieces; add remaining ingredients.

Dressing:

5 tablespoons raspberry wine vinegar
¼ teaspoon dry mustard
2 tablespoons sugar
2 tablespoons dried parsley

Garlic powder to taste
½ cup sour cream
1 teaspoon cracked pepper
1 cup oil

Combine first seven dressing ingredients in a quart jar; shake well. Add oil; shake again and refrigerate. Shake dressing several times before serving to guarantee blending of oil and vinegar. Pour over salad, toss gently, and serve.

Salad can be made several hours in advance. Combine 1 cup of dressing with avocados and scallions in bottom of salad bowl. Add lettuce and place other ingredients on top of the lettuce. Refrigerate and toss when ready to serve. The dressing can be made and stored for weeks in the refrigerator.

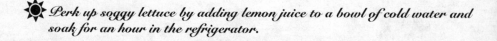

Perk up soggy lettuce by adding lemon juice to a bowl of cold water and soak for an hour in the refrigerator.

SUMMER SPINACH SALAD

Serves: 8

Salad:

2 bunches spinach, washed and cut

1 (15 oz.) can mandarin oranges, drained

1 pint strawberries, halved

1 red onion, sliced thin

Pitted black olives, optional

Wash all ingredients, dry well. Mix all salad ingredients and chill.

Dressing:

½ cup sugar

2 tablespoons sesame seeds

1 tablespoon poppy seeds

1½ tablespoons minced onion

½ teaspoon Worcestershire sauce

½ cup olive or vegetable oil

¼ cider or balsamic vinegar

¼ teaspoon paprika

Combine dressing ingredients. Refrigerate until serving time. Mix well and pour over salad just before serving.

SPINACH SALAD

Serves: 14

Salad:

1 pound fresh spinach, washed and torn into bite-size pieces

1 (14 oz.) can hearts of palm

1 (14 oz.) can artichoke hearts, drained and diced

6 slices bacon; cooked and crumbled

4 green onions, chopped

2-3 tomatoes, halved

Slivered almonds, lightly toasted

Combine all salad ingredients in a large salad bowl. Toss, cover, and chill until serving time.

Dressing:

1 small onion, minced

¾ cup sugar

½ cup vinegar

⅓ cup ketchup

¼ cup salad oil

2 tablespoons Worcestershire sauce

To prepare dressing, combine all ingredients in blender, blending well and chilling before serving. Dressing makes approximately 2 cups.

 To peel oranges and grapefruit without the white membrane adhering to the sections, soak the fruit in hot water for 5 minutes before peeling.

NECTARINE RED ONION SALAD

Serves: 4

Salad:

1 avocado, sliced
Seasoned salt to taste
¼ red onion, thinly sliced
2 nectarines, unpeeled and cut
into small wedges

1 large head of red tip lettuce
(or ½ red tip and 1 head of
Boston lettuce)

Toss avocado, onion, nectarines, and lettuce in a large bowl.

Dressing:

½ cup olive oil, extra virgin
1 tablespoon balsamic vinegar
1 tablespoon raspberry wine
vinegar

½ teaspoon salt
Cracked black pepper to taste

Combine olive oil, balsamic vinegar, raspberry vinegar, salt, and black pepper
in a jar. Shake well. When ready to serve, pour dressing over salad. Diced grilled
chicken may be added if desired.

ASPARAGUS VINAIGRETTE

Serves: 4

¾ pound fresh asparagus spears
¾ cup Italian salad dressing
2 tablespoons pimiento-stuffed
green olives, finely chopped

1 hard boiled egg, finely
chopped
2 small tomatoes, chilled
Lettuce leaves

Steam asparagus spears for 10-15 minutes, until crisp-tender. Combine salad
dressing, chopped olives, and eggs; mix well. Arrange asparagus spears in a
shallow dish. Pour salad dressing mixture over asparagus. Cover and refrigerate
for several hours or overnight, spooning dressing over asparagus occasionally.
To serve, drain asparagus, reserving dressing. Slice tomatoes. On each of 4
salad plates, arrange a few spears atop lettuce; top each salad with a few
tomato slices. Spoon some of the reserved dressing over each salad.

BROCCOLI SALAD

Serves: 4

Salad:
- 3-4 cups broccoli
- 10 slices bacon
- ½ cup diced white onion
- 1 cup raisins
- 1 cup salted sunflower seeds

Wash and drain broccoli. Cut stem off; break into bite size pieces. Cook bacon in microwave until crispy. Crumble and add to broccoli with onion, raisins, and sunflower seeds.

Dressing:
- 1 cup mayonnaise
- 2 tablespoons vinegar
- ⅓ cup sugar

Mix dressing ingredients together; toss into salad.

BROCCOLI GRAPE SALAD

Serves: 10-12

Salad:
- 1 bunch broccoli, cut in bite size pieces
- 1 cup chopped celery
- ¼ cup chopped green onions, including tops
- 1 cup seedless green grapes, sliced in half
- 1 cup seedless red grapes, sliced in half
- 1 cup slivered almonds
- ½ pound bacon

Cook bacon until crisp; drain, and crumble. Mix with other ingredients.

Dressing:
- ¼ cup sugar
- 1 cup mayonnaise
- 1 tablespoon vinegar

OR
- 1 (8 oz.) bottle poppy seed dressing

Whisk sugar, mayonnaise, and vinegar together; pour over salad and toss one hour before serving or use bottled poppy seed dressing.

97

SWEET AND SPICY CARROTS

Serves: 8-10

2 pounds baby carrots, scraped

Dressing:
¾ cup sugar
¾ cup vegetable oil
1 tablespoon
Worcestershire sauce
1 medium onion, chopped

1 (10 oz.) can tomato soup,
undiluted
½ cup vinegar
1 green bell pepper, chopped
1 tablespoon mustard

Cook carrots very slightly. Marinate in dressing several hours or overnight.
Serve cold.

MARINATED CORN

Serves: 8

Salad:
2 (16 oz.) cans white shoepeg
corn, drained
½ cup chopped green bell
pepper

½ cup chopped onion
2 stalks celery, chopped
1 (2 oz.) jar pimiento, diced

Marinade:
½ cup vegetable oil
½ cup vinegar
½ cup sugar

1 teaspoon salt
½ teaspoon pepper

Combine all ingredients and marinate overnight. Drain before serving.

 To prevent a vegetable salad from becoming soggy when it has to stand for a few hours, place a saucer upside down on the bottom of the bowl before filling it with salad. The moisture will run underneath and the salad will remain fresh and crisp.

CORNBREAD SALAD

Serves: 6-8

1 (16 oz.) package jalapeño
cornbread mix
1 small onion, chopped
1 green bell pepper, chopped
1 fresh tomato, chopped
½ cup mayonnaise

2 tablespoons prepared
mustard
½ teaspoon salt
½ teaspoon pepper
1 teaspoon sugar

Prepare cornbread according to package directions; cool. Crumble cornbread after it cools. Mix with remaining ingredients and chill for several hours or overnight.

ROASTED EGGPLANT SALAD

Serves: 8

1-1½ cups olive oil
2 cups diced red bell pepper
2 cups diced green bell pepper
2 cups diced white onion
6 cups diced eggplant
1 cup canned diced tomatoes in
juice

2 tablespoons capers
¼ cup red wine vinegar
¼ cup sugar
¼ cup minced cilantro
1 tablespoon minced anchovies
(optional)
1 tablespoon minced garlic

Heat a few tablespoons oil in heavy skillet on high flame until smoking. Mix together peppers and onions. Carefully add mixture to oil in small batches; sauté until browned but still crisp, about 3-4 minutes, adding more oil as needed. (Use enough oil to cover bottom of skillet; do not clean skillet between batches.) Remove to mixing bowl. Repeat procedure with eggplant; add to peppers.
In heavy saucepan, combine tomatoes, capers, vinegar, and sugar. Simmer slowly until syrupy, about 10 minutes. Add to eggplant mixture.
In a food processor, puree remaining ingredients to consistency of paste, scraping sides of bowl frequently. Mix well with vegetables, and refrigerate overnight. Serve chilled or at room temperature.

GAZPACHO SALAD MOLD

Serves: 6-8

1 envelope unflavored gelatin
1⅔ cups tomato juice
2 tablespoons red wine vinegar
1 large tomato, peeled, seeded, and chopped
1 cucumber, peeled and chopped
1 (4 oz.) can green chilies, drained, seeded, chopped

¼ cup sliced green onions
1 clove garlic, minced
¾ teaspoon salt
⅛ teaspoon ground black pepper
Pinch of sugar
Crisp salad greens
Sour cream to garnish

1 quart mold, or 4-6 individual molds

Soften gelatin in ¼ cup tomato juice. Heat remaining juice in medium saucepan. Add softened gelatin; stir until dissolved. Add all other ingredients except greens and sour cream. Pour into 1 quart mold or individual molds. Chill overnight, or until set. Unmold onto salad greens; serve with dollop of sour cream.

SPICY POTATO SALAD

Serves: 8-10

6-8 medium potatoes
½ cup mayonnaise
1 cup sour cream
1 teaspoon salt
1 tablespoon horseradish
½ cup chopped onions
Parsley garnish

Optional:
Bacon Bits
Pepper to taste
½ cup chopped green bell pepper

Boil unpeeled whole potatoes. Peel and slice. Layer ¼ of potatoes in dish. Mix remaining ingredients; divide in fourths. Spread ¼ of creamy mixture over potatoes in dish; continue to alternate layers (potatoes, sauce, etc.) Chill well before serving.

 One tablespoon of vinegar with one quart of water when washing green or any leafy vegetable will guarantee clean vegetables in half the washing time.

MARINATED SLAW

Serves: 8-10

Hot Dressing:

1 teaspoon celery salt
1 teaspoon sugar
1 teaspoon dry mustard

1 teaspoon salt
1 cup cider vinegar
1 cup oil

Combine celery salt, sugar, mustard, salt, and vinegar in saucepan. Bring to a rolling boil. Add oil, stirring, and again return to a rolling boil.

Slaw:

¾ cup sugar
1 large head cabbage, shredded

1 medium to large red onion,
thinly sliced

Stir sugar into cabbage. Place half of cabbage in large bowl. Cover with onion slices. Top with remaining cabbage. Pour boiling Hot Dressing over slowly. Do not stir. Cover and refrigerate at once. Chill 24 hours. Stir well before serving.

VERMICELLI SALAD

Serves: 8

10 ounces vermicelli, cooked and
drained
1 large purple onion, chopped
1 large tomato, diced
1 cucumber, diced

1 bell pepper, diced
1 zucchini, diced
1 (2.62 oz.) bottle McCormick
Salad Supreme Seasoning
1 (16 oz.) bottle Italian dressing

Toss vermicelli, onion, tomato, cucumber, bell pepper, and zucchini. Sprinkle with bottle of Salad Supreme. Add Italian dressing. Seal tightly and marinate overnight.

 To hasten the ripening of garden tomatoes or avocados, put them in a brown paper bag, close the bag, and leave at room temperature for a few days.

TOMATO ASPIC

Serves: 8-10

2 (3 oz.) boxes of lemon gelatin
1½ cups clamato juice
¾ cup chopped celery
½ cup chopped green bell
  pepper

1 (7 oz.) can salsa or Rotel
1 (8 oz.) can tomato sauce
½ cup chopped onion

8 x 8 glass dish

Dissolve gelatin in hot clamato juice. Add remaining ingredients; pour mixture into dish. Refrigerate until firm, 2-3 hours.

FRUIT MEDLEY

Serves: 15

1 (14 oz.) can sweetened
  condensed milk
1 (20 oz.) can cherry pie filling
1 (12 oz.) carton whipped
  topping
1 cup flaked coconut

1 cup chopped apple
1 cup sliced banana
1 cup chopped pecans
1 cup green grapes
1 cup pineapple chunks, drained

Stir all ingredients together in large bowl. Refrigerate. This will keep two weeks without the apples and bananas. (Add them when ready to serve.) The base will freeze well.

APPLE SALAD

Serves: 4-6

1-2 apples, cut in bite size pieces
1 cup miniature marshmallows
1 cup chopped nuts, your
  choice

2 rounded tablespoons salad
  dressing
2-3 tablespoons milk

Toss apples, marshmallows, and nuts together. Add milk a little at a time to salad dressing until creamy and desired thickness. Pour over salad and mix well. Cover and refrigerate until served.

Dissolve gelatin in the microwave. Measure liquid in a measuring cup, add gelatin and heat. There will be less stirring to dissolve the gelatin.

CONGEALED CRANBERRY SALAD

Serves: 15-20

1 (6 oz.) box raspberry gelatin
3 cups boiling water
1 pound fresh cranberries
2 cups sugar

1 (20 oz.) can crushed
   pineapple, drained
1½ cups chopped pecans
Grated orange rind (optional)

3 quart glass bowl or individual molds

In large mixing bowl, combine gelatin and water. Set aside to cool. Wash and drain cranberries and chop in food processor. Add sugar and mix well. Fold all ingredients into cooled gelatin. Pour into desired container and refrigerate 4 to 6 hours.

LUSCIOUS STRAWBERRY SALAD

Serves: 6-8

2 (3 oz.) boxes strawberry
   gelatin
2 cups boiling water
1 (16 oz.) package frozen
   strawberries

3 bananas, sliced
1 (8 oz.) carton sour cream

6 cup mold or 9 inch square dish

Dissolve gelatin in water; add strawberries; stir until strawberries are slightly thawed. Add bananas. Pour half of mixture into mold or dish. Chill until firm. Spread sour cream over firm gelatin. Add remaining gelatin over sour cream. Return to refrigerator and chill until set.

CRANBERRY SALAD

Serves: 12-15

1 (16 oz.) can whole cranberry
   sauce
3 medium bananas, mashed
2 cups miniature marshmallows

1 (20 oz.) can crushed
   pineapple
1 cup chopped pecans
2 cups whipped topping

In large mixing bowl, combine all ingredients. Mix with spoon and pour into glass dish. Place in freezer until firm.

If you wet the dish on which the gelatin is to be unmolded, it can be moved around until centered.

103

MANGO MOLD

Serves: 8-10

1 (20 oz.) can mangos
1 (8 oz.) package cream cheese
1 (6 oz.) box lemon gelatin

1 envelope unflavored gelatin
Water

6 cup ring mold

Drain mangos, reserving liquid. Add enough water to liquid to make one cup. Add both packages of gelatin to water in saucepan. Heat just to boiling. Blend mangos and cream cheese together in blender. Combine liquid and mango mixture and pour into mold. Chill until set.

Use 4 large fresh mangos, 2 envelopes unflavored gelatin, and ½ cup sugar. Follow above directions, squeezing mangos with spoon or fork to remove the juice.

GRAPEFRUIT WEDGE SALAD

Serves: 12

3 large Ruby Red grapefruit
1 (16 oz.) can apricots
1 (3 oz.) box lemon gelatin

1 tablespoon sugar
½ cup chopped pecans

Cut grapefruit in halves. Remove pulp from rind so that rind remains intact. Cut enough segments to fill one cup. Squeeze enough juice for 1 cup. Drain juice from apricots; reserve. Add grapefruit juice to apricot juice, which will be two cups total. Heat juice; add to gelatin and sugar. Cool slightly. Mash apricots. Add apricots, grapefruit segments, and pecans to gelatin mixture. Fill empty rind halves with mixture; set in pan in refrigerator to congeal.

To serve, cut each half in quarters. Serve on fruit leaves or English ivy leaves. Poppy seed dressing poured over each quarter is a nice addition.

FIESTA FRUIT SALAD

Serves: 8-12

1 (15 oz.) can pineapple chunks
1 (15 oz.) can pears, cut into chunks
1 (6 oz.) jar red cherries, sliced in half

1 (11 oz.) can mandarin oranges
1 cup miniature marshmallows, optional
1 (8 oz.) carton sour cream

Drain all fruit very well on paper towels. In bowl, add sour cream to fruit and stir. Chill overnight.

CHICKEN SALAD

Serves: 8-10

4 cups cooked and diced chicken breasts
1 cup diced celery
1 cup diced tart apple
¼ cup sliced scallions
½ cup chopped walnuts or pecans

¼ cup snipped fresh dill
¼ cup chopped fresh parsley
1 cup mayonnaise
1 tablespoon lemon or lime juice
½ teaspoon salt
Ground white pepper to taste

Mix chicken, celery, apple, scallions, walnuts, dill, parsley, mayonnaise, lemon juice, salt, and pepper well. Chill.

HOT CHICKEN SALAD

Serves: 8-10

1 (10½ oz.) can cream of chicken soup
¾ cup mayonnaise
3 tablespoons lemon juice
½ teaspoon salt
¼ teaspoon black pepper
2 cups diced cooked chicken
1 medium onion, diced

1 cup diced celery
1 (8 oz.) can water chestnuts, sliced
1 cup slivered almonds, slightly toasted and buttered
2 cups crushed sour cream and onion potato chips

Preheat oven to 450°.   13 x 9 baking dish

Grease baking dish. Blend together soup, mayonnaise, lemon juice, salt, and pepper. Add chicken, celery, onion, almonds, and water chestnuts. Mix well. Put into buttered baking dish, cover with potato chips, and bake 20 minutes, or until sauce is bubbly.

 To toast nuts or seeds, microwave 1/4 cup of nuts or seeds and 1 teaspoon of butter on high power for 5 minutes, stirring once after 2 minutes.

ELEGANT LUNCHEON SALAD

Serves: 8

4 chicken breasts, boned, boiled
 with 1 teaspoon salt
1 (4¾ oz.) package Chicken
 Rice-A-Roni, cooked and
 cooled
1 (6 oz.) can pitted ripe olives
1 (6 oz.) jar marinated
 artichoke hearts, undrained
 and chopped

1 bunch green onions, sliced
1 cup mayonnaise or Lite
 Miracle Whip
 Salt
 Pepper
½ cup slivered almonds, toasted

Mix chicken, Rice-A-Roni, olives, artichoke hearts, onions, mayonnaise, salt, pepper, and almonds well. Flavor improves after 24 hours.

SHRIMP SALAD, PADRE STYLE

Serves: 4

1 cup cooked rice
1 cup cooked shrimp, shelled,
 deveined, and diced
⅛ teaspoon salt
1 tablespoon fresh lemon juice
¼ green bell pepper, slivered
1 tablespoon minced onion
 (white, yellow, or green)

1 tablespoon chopped black
 olives
¾ cup diced raw cauliflower
2 tablespoons French dressing
 (may use low calorie)
 Dash of black pepper
⅓ cup mayonnaise

Toss all ingredients together; chill, and serve on a bed of lettuce leaves.

Lower salt content of olives by soaking them in cold water in the refrigerator for several hours or overnight.

ORIENTAL CHICKEN SALAD

Serves: 6

Marinade:

⅓ cup sherry
¼ cup soy sauce
1 (2 inch) piece of ginger, thinly
  sliced

2 large whole chicken breasts,
  boned
Salt
Cornstarch

Combine sherry, soy sauce, and ginger in a medium bowl. Add chicken; cover and refrigerate several hours or overnight, turning once. Drain and pat dry. Lightly salt chicken and coat pieces in cornstarch. Refrigerate at least one hour to allow cornstarch to set.

Salad:

3 cups iceberg lettuce,
  shredded (½ head)
6 green onions, sliced
  diagonally
1 bunch fresh cilantro, chopped

½ cup sliced almonds, toasted
¼ cup sesame seeds, toasted
2-3 ounces fried rice noodles
Oil for deep frying
Alfalfa sprouts

In a large skillet, heat about two inches oil to 375°. Add chicken and deep fry until golden brown and crisp. Drain on paper towels and let cool. Slice chicken, with skin on, into bite size pieces and place in bowl. Add lettuce, onions, and cilantro. Add salad dressing and mix gently. Sprinkle with almonds and sesame seeds; garnish with alfalfa sprouts.

Dressing:

2 teaspoons dry mustard
  powder
1 teaspoon grated fresh ginger
2 teaspoons sugar
2 tablespoons lemon juice

½ cup salad oil
2 teaspoons sesame oil
1 teaspoon hot pepper oil
1½ tablespoons soy sauce

In a blender, mix mustard, ginger, and sugar. Add lemon juice and then slowly pour in the oils and soy sauce.

 To remove the core from a head of lettuce, hit the core end once against the counter sharply. The core will loosen and pull out easily.

INDONESIAN NOODLES

Serves: 16-20

Pasta:

3 (9 oz.) packages angel hair pasta or Oriental rice stick noodles

4 pounds small shrimp (35-40 per pound)

1 bunch scallions cut into 2 inch lengths, then into thin strips

1 pound snow peas

1 red bell pepper, seeded and cut into thin strips

1 yellow bell pepper, seeded and cut into thin strips

1 (8 oz.) can water chestnuts, drained and thinly sliced

2 (8 oz.) cans sliced bamboo shoots, drained

1 (7¼ oz.) jar pickled baby corn, drained and cut lengthwise into quarters

½ cup chopped fresh cilantro

¾ cup salted peanuts

Prepare pasta as per directions on the package; place in a large mixing bowl. Cook shrimp, that have been peeled and deveined, until opaque. Blanch snow peas, with strings removed, in boiling water for 30 seconds; then rinse under cold running water. Cut in half lengthwise. Add the scallions, snow peas, bell peppers, water chestnuts, bamboo shoots, baby corn, cilantro, shrimp, and peanuts to pasta; toss to combine.

Dressing:

¼ cup Oriental sesame oil

¾ cup soy sauce

3½ cups mayonnaise

Hot chili oil to taste

Whisk the sesame oil, soy sauce, and mayonnaise together in a small mixing bowl. Pour the dressing over the noodles and mix thoroughly with your hands to ensure that the salad is evenly coated. Season with drops of hot chili oil to taste. Cover salad and refrigerate for several hours before serving.

Breads

Refer to Mexican Cuisine section for additional recipes in this category.

Teen Court

In conjunction with the McAllen Police Department, our Teen Court program encourages young people, 10-17 years of age, to be responsible for their actions. Teens who have been charged with a first time misdemeanor are tried by a jury of their peers and are given community service in lieu of paying fines. Our volunteers oversee the entire program which also includes teen attorneys, court clerks, jurors, and bailiffs. Local judges give their time to ensure the success of the program. Our goal is to redirect teen behavior before it leads to a lifetime of criminal activity.

 This recipe is pictured on the previous page.

ANGEL ROLLS

Yield: 20-25 rolls

5 cups flour
¼ cup sugar
1 tablespoon baking powder
1 teaspoon baking soda
1 teaspoon salt

1 cup shortening
1 package dry yeast
¼ cup warm water
2 cups buttermilk
½ cup butter, melted

Preheat oven to 400°.   Cookie Sheet

Sift together flour, sugar, baking powder, baking soda, and salt; cut in shortening. Dissolve yeast in warm water and add with buttermilk to dry ingredients. Mix well. Knead on a floured surface until smooth and satiny; roll out on a floured board. Cut with a round cutter, dip in melted butter, fold in half, and place on cookie sheet. Bake for 12-15 minutes.

Put cookie sheet of unbaked rolls in refrigerator to bake later or put in freezer; when frozen, put in plastic bag.

BRAN ROLLS

Yield: 24 rolls

¾ cup whole bran cereal
⅓ cup sugar
1½ teaspoons salt
½ cup margarine
½ cup boiling water

½ cup warm water (105°-115°)
2 packages active dry yeast
1 egg, beaten
3¼-3¾ cups unsifted flour
Melted margarine

Preheat oven to 375°.   2 (½-inch) muffin pans or 2 (8-inch) round cake pans

Grease muffin or round cake pans. Combine cereal, sugar, salt, and margarine in a bowl. Add boiling water; stir until margarine is melted. Set aside to cool to lukewarm. Measure warm water into large bowl; sprinkle in yeast; stir until dissolved. Add lukewarm cereal mixture, egg, and enough flour to make a stiff dough. Turn out onto lightly floured board; knead until smooth and elastic, about 8-10 minutes. Place in greased bowl, turning to grease top. Cover; let rise in warm place, free from drafts, until doubled in bulk, about 1 hour. Punch dough down; divide in half. Divide each half into 12 equal pieces. Form each piece into a smooth round ball. Place in muffin pans or cake pans. Brush rolls with melted margarine. Cover; let rise in warm place, free from draft, until doubled in bulk, about 1 hour. Bake 20-25 minutes, or until done. Remove from pans and cool on wire racks.

CARAWAY BRAID

Yield: 2 loaves

2 packages dry yeast
¼ cup warm water
¼ teaspoon sugar
2 cups milk
4 teaspoons caraway seed
1 teaspoon anise seed
¼ cup sugar

⅓ cup shortening
1 tablespoon salt
3 eggs, beaten
7 cups sifted flour (use more if needed)
1 tablespoon butter, melted
1 egg, beaten

Preheat oven to 375° during second rising.

Sprinkle yeast and sugar over warm water; let soften. Heat milk and seeds to boiling. Remove from heat; pour into large bowl. Add sugar, shortening, and salt. Stir until blended. Beat eggs in second bowl; add to milk mixture. When cooled to lukewarm, add yeast mixture. Beat in flour, 2 cups at a time, until smooth. Stir with spoon until the dough forms a ball; turn out on floured board. Knead until satiny. Place in greased bowl, turning over once to grease top of dough. Cover and place in warm area to rise (approximately 2 hours). When dough has doubled in bulk, turn out on floured board. Divide into 2 parts. Cut each half in 3 portions. Roll each third into a rope 14 inches long. Make an "X" with two strips. Place third down the center. Braid from center to both ends. Tuck ends under. Brush with melted butter. Cover with wax paper and then with cloth. Place in warm place until doubled in bulk. Brush with beaten egg. Bake 40 minutes or until well browned. Baked loaf sounds hollow when rapped with knuckles. Remove to rack to cool.

Freezes well. Do not double recipe.

DOUGHNUTS

Yield: 1 dozen

3 eggs
1 cup ricotta cheese
1 cup flour
¼ cup sugar

3 teaspoons baking powder
Cooking oil
Powdered sugar

Beat eggs with ricotta cheese until smooth. Mix flour, sugar, and baking powder; add to egg mixture. Drop by spoonsful into hot oil. When lightly browned remove from oil and roll in powdered sugar.

CHALLA (EGG) BREAD

Yield: 2 loaves

¾ cup sugar, divided
2 packages dry yeast
¾-1 cup warm water, (105°)
1 cup oil
6 eggs, well beaten

1 tablespoon salt
8 cups flour, sifted
1 egg yolk
2 tablespoons water
Poppy seeds

Preheat oven to 375° during third rising.

Mix together approximately 2 tablespoons sugar, yeast, and 1 cup water. Let stand about 10 minutes, until it bubbles. Add oil, eggs, salt, rest of sugar, and 2 cups flour. Beat with mixer. Cover. Let rise 45 minutes. Add remaining flour and let rise for 2 hours. Sprinkle wooden pastry board and hands with flour.
Punch down dough and separate into 6 pieces. Roll these pieces into long strands. Braid 3 strands together tightly for each loaf. Place the 2 loaves on a lightly greased baking sheet. Cover and let rise in a warm place until double in size (1 hour). Brush with a mixture of 1 egg yolk and 2 tablespoons water. Sprinkle with poppy seeds. Bake 30-40 minutes or until golden brown.

CHEESE BREAD

Yield: 2 loaves

2 cups milk
2½ tablespoons margarine
⅛ cup light corn syrup
1 tablespoon brown sugar
5-6 cups flour
1 package dry yeast

2 teaspoons salt
1 teaspoon dried parsley
1 teaspoon garlic powder
1 (10 oz.) package extra sharp
Cheddar cheese, grated

Preheat oven to 375°.   2 (9 x 5) loaf pans

Grease loaf pans. In saucepan bring milk, margarine, corn syrup, and brown sugar to a boil; cool to lukewarm. In a large bowl, mix 2 cups flour, yeast, salt, parsley, and garlic powder; add milk mixture to the dry ingredients, then cheese. Mix well; adding more flour until the dough is soft enough to knead (about 5 minutes). Place on a lightly floured surface and knead until very smooth. Put in a greased bowl; turn to grease top and cover. Let rise until doubled in size (1½-2 hours). Punch down, let rest for 10 minutes. Divide dough; shape into loaves (be careful not to overhandle dough) and allow to double in size, about 45 minutes. Bake for 35-45 minutes; remove from pans at once.

DINNER ROLLS

Yield: 4 dozen

1 cup shortening
¾ cup sugar
2 teaspoons salt
2 yeast cakes or 2 packages
   dry yeast

1 cup lukewarm water
1 cup boiling water
2 eggs, beaten
6 cups unsifted flour
½ cup butter, melted

Preheat oven to 425° before baking.   Jelly roll pan

Grease jelly roll pan. Mix shortening, sugar, and salt in a large bowl with electric mixer. In a small bowl, dissolve yeast in warm water. Add boiling water to shortening mixture; add beaten eggs, yeast, and flour. Mix thoroughly. Chill dough at least 4 hours or overnight. Roll out on floured surface and cut out with round cutter. Dip one side into melted butter and fold butter side inward. Place rolls in pan, slightly touching. Bake 10 minutes or until brown on top.

POTATO ROLLS

Yield: 3 dozen

1 package dry yeast
1 teaspoon sugar
⅓ cup warm water
¾ cup shortening
½ cup sugar
2 teaspoons salt

½ cup scalded milk
½ cup hot potato water
1 cup hot mashed potatoes
3 eggs, slightly beaten
6 cups flour

Preheat oven to 425°.   Muffin tins

Grease muffin tins. Dissolve yeast in sugar and warm water. In a separate large bowl, combine shortening, sugar, salt, milk, potato water, and mashed potatoes; let cool to lukewarm. Add yeast mixture and eggs; slowly add flour, eventually mixing by hand or using dough hook on mixer. Place in greased bowl and cover. At this point the dough can be refrigerated for up to 4 weeks. When ready to use, shape into muffin tins by rolling 3 balls for each muffin. Let rise 2 hours, bake for 15 minutes.

 Put a small pan with boiling water in the oven with the rising bread dough. It cuts the rising time in half.

FRENCH BREAD

Yield: 2 loaves

1 package dry yeast
1 tablespoon sugar
2 teaspoons salt
2 cups warm water

4 cups flour
Melted butter
1 egg

Preheat oven to 450° during third rising.   Baguette pan

In a large bowl dissolve yeast, sugar, and salt in warm water. Let stand 10 minutes. Stir in flour and turn onto floured surface. Add more flour if necessary and knead for 10 minutes, or until smooth and elastic. Put in clean bowl, cover with plastic wrap and let rise 45 minutes in a warm place.

Loosen dough from sides of bowl with a rubber spatula and turn out. Gently pat flat, dust with a little flour, and fold to form a half circle, and then again to form a quarter circle. Return to bowl and let rise for another 30 minutes. (The rising develops the full flavor and the ultimate lightness.)

Turn dough out onto a lightly floured surface, pat flat and fold in half. Cut into two equal pieces (if only 1 pan is being used, put half the dough back in bowl for another rising), form into balls and let rest 5 minutes. Flatten each ball into an oval and roll up like a jelly roll. Form into a loaf almost the length of the pan by rolling back and forth.

Line baguette pan with a pastry cloth or other coarsely woven cloth and place formed dough in furrows. Cover with a clean towel and let rise again until almost double. Pull the pan out from under the cloth. Brush the pan with butter. Gently flip the loaves onto the pans. Use the cloth to help, pulling the loaf towards the pan. Brush the loaves with well-beaten egg mixed with a little water, and slash the loaves diagonally with a knife or razor blade. Bake for 20 to 25 minutes.

HERBED GARLIC BREAD

Yield: 1 loaf

1 loaf French bread
½ cup softened butter or
   margarine
1 teaspoon parsley

¼ teaspoon oregano
¼ teaspoon dill
1 clove garlic, pressed

Preheat oven to 400°.   Cookie sheet

Slice loaf into one inch diagonals. Blend remaining ingredients together. Spread mixture on slices. Reassemble loaf, wrap in foil, and bake for 10-15 minutes. Easy!

Top with freshly grated Parmesan cheese, if desired.

REFRIGERATOR ROLLS

Yield: 4 dozen rolls

2½ cups very hot water
⅔ cup sugar
¼ cup shortening
3½ teaspoons salt
3 eggs

1 package dry yeast
8 cups flour
10 ounces mild Cheddar cheese, shredded (optional for cheese rolls)

Preheat oven to 400° before baking.   13 x 9 pan

Grease pan. Pour 2½ cups very hot water over sugar, shortening, and salt. Let cool to lukewarm. Add eggs and yeast. Mix in cheese, if desired. Stir in 4 cups of flour, using electric mixer. Remove from mixer; continue adding 4 cups of flour with a spoon or your hands. Refrigerate overnight. Grease hands to shape rolls. Let rise 3-5 hours. Bake 20 minutes.

Best if dough is prepared 2 to 3 days in advance. May be kept in refrigerator for 1 week.

WHEAT GERM BRAID BREAD

Yield: 3 loaves

½ cup shortening
1-2 teaspoons salt
½ cup sugar
1 cup hot milk
2 eggs, beaten

2 packages dry yeast
½ cup warm water (105°-115°)
1 cup wheat germ
6½-7½ cups flour, divided

Preheat oven to 325° during second rising.   2-3 cookie sheets

Grease cookie sheets. Place shortening, salt, and sugar in large bowl and add hot milk. Stir to soften; cool. Beat in eggs. Soften yeast in warm water; stir into shortening mixture. Add wheat germ and 3½ cups flour. Beat until smooth; add enough flour to make a soft dough. Turn out on floured surface, and knead until smooth. Place in greased bowl, turn once to grease surface. Cover; let rise until doubled in size. Punch down; let rest 10 minutes. Divide into 12 or 15 pieces (3 for each loaf). Shape each piece into 12-inch lengths and braid into loaves. Place on cookie sheets; let rise until double. Brush with beaten egg, and bake for 15-20 minutes.

By letting your bread rise in a plastic bag, you can punch down hard and knead it vigorously, since dough does not dry out.

WHITE BREAD

Yield: 2 loaves

2½ cups warm water (105°-115°)
2 packages dry yeast
½ cup instant non-fat dry milk
2 tablespoons sugar

1 tablespoon salt
⅓ cup vegetable oil
7 cups flour, divided

Preheat oven to 400° (or 350° for glass pan).

1 (9 x 5) and 1 (7½ x 3½) loaf pan or 2 (8½ x 4½) pans

Grease loaf pans. Place warm water in a large mixing bowl; sprinkle yeast over top. Add dry milk, sugar, salt, oil, and about 3 cups flour. On low speed of electric mixer, blend well. Scrape sides of mixing bowl clean. Increase mixer speed to medium for 3 additional minutes.

Remove bowl from mixer; using a wooden spoon, stir in remaining flour. The dough will become stiff, so don't be afraid to use your hands to mix flour in.

Place dough on a lightly floured board, cover with a towel, and let rest for 15 minutes.

Then, knead dough about 1 minute — only until smooth. Divide dough into 2 pieces using serrated knife. Shape into balls; let rest for 4 minutes under a towel. Form a loaf by pressing a ball of dough into a flat oval (roughly the length of the baking pan you wish to use). Fold the oval in half lengthwise, pinch the seam tightly to seal, tuck under the ends, and place seam down in the pan. (Repeat for other loaf.)

Place the pans in a warm (draft free) place (80°-85°), cover with wax paper, and let rise about 1 hour.

Bake for 10 minutes; reduce heat to 350° (325° for glass pan); continue baking for another 35 minutes. Halfway through the baking, turn pans. Turn again 10 minutes before cooking process is finished.

Turn loaves from pans onto wire racks to cool.

 A heated knife blade will cut through fresh bread more easily.

HONEY WHOLE WHEAT BREAD
Yield: 2 loaves

3 cups whole wheat flour
½ cup nonfat dry milk
1 tablespoon salt
2 packages dry yeast
3 cups water

½ cup honey
2 tablespoons vegetable oil
2 cups whole wheat flour
4-4½ cups all-purpose flour

Preheat oven to 375° during second rising. 2 (9 x 5) loaf pans

Lightly grease loaf pans. Combine 3 cups whole wheat flour, nonfat dry milk, salt, and yeast in large mixing bowl. Heat water, honey, and vegetable oil in saucepan until warm. Pour over flour mixture. Blend with electric mixer on low speed 1 minute, medium speed 2 minutes. Stir in by hand remaining whole wheat and all-purpose flours. Knead 5 minutes. Place dough in large greased bowl; let rise until double in size. Punch down, then shape into 2 loaves. Place in prepared pans. Cover. Let rise until double in size. Bake for 40-45 minutes.

WHOLE WHEAT ROLLS
Yield: 36-40 rolls

1 package dry yeast or 1 yeast
  cake
2 tablespoons lukewarm water
¼ cup butter or margarine
1 cup scalded milk

⅓ cup honey
1½ teaspoons salt
2 eggs, beaten
4½ cups whole wheat flour or
  unbleached flour

Preheat oven to 350° during second rising. Cookie sheet

Grease cookie sheet. Dissolve yeast in water. Melt butter in scalded milk. Add honey and salt to the milk mixture. Let cool. Add yeast mixture and eggs, whisk to blend. Add flour, a little at a time. Let rise until double in bulk. Knead a little on a floured board. Roll out and cut with biscuit cutter. Place on cookie sheet; let rise 1 hour. Bake for 15-20 minutes. Freezes well.

 To scald milk, cook 1 cup milk for 2-2 1/2 minutes, stirring once each minute.

WHOLE WHEAT BUTTERHORNS
Yield: 18-24 rolls

2¾ cups all-purpose flour, divided
2 packages dry yeast
1¾ cups water
⅓ cup brown sugar, firmly
    packed

½ cup butter or margarine,
    divided
2 tablespoons honey
2 teaspoons salt
2 cups whole wheat flour

Preheat oven to 400° during second rising.   2 cookie sheets

Grease cookie sheets. Combine 1½ cups flour and yeast in large mixing bowl. Heat water, brown sugar, 3 tablespoons butter or margarine, honey, and salt to 120°-130°; add to flour mixture. Beat with electric mixer on low for 30 seconds; increase speed to high and continue beating for 3 minutes. Stir in whole wheat flour and enough all-purpose flour to form a soft dough. Place dough on a lightly floured surface; knead until smooth and elastic, about 6-8 minutes. Place in a large greased bowl; cover and allow to rise in a warm place until double in size, approximately 1½ hours. Punch dough down and divide into thirds. Shape each third into a ball; cover and let rest 10 minutes. On lightly floured surface, roll balls into 3 (12 inch) circles. Cut each circle into 6-8 wedges. Roll triangular wedges into crescent shapes starting at wide end. Place on cookie sheets; cover and let rise until doubled, approximately 1 hour. Melt remaining butter and brush on each crescent. Bake for 10-15 minutes until golden brown. Brush again with butter while hot.

FRESH APPLE BREAD
Serves: 8

1 cup sugar
½ cup shortening
2 eggs
1 teaspoon vanilla extract
2 cups flour
1 teaspoon baking powder

1 teaspoon baking soda
½ teaspoon salt
2 cups chopped apples
½ cup chopped nuts
1 tablespoon sugar
¼ teaspoon cinnamon

Preheat oven to 350°.   9 x 5 loaf pan

Grease and flour loaf pan. Mix sugar, shortening, eggs, and vanilla. Stir in flour, baking powder, baking soda, and salt. Stir until smooth; add apples and nuts. Pour into prepared loaf pan. Mix sugar and cinnamon together. Sprinkle over batter. Bake for 50-60 minutes.

APRICOT BREAD

Yield: 1 loaf

½ cup chopped dried apricots
½ cup seedless raisins
Grated rind of 1 large orange
Juice of 1 large orange
Boiling water
1 teaspoon baking soda
¾ cup sugar

2 tablespoons unsalted butter, melted
1 teaspoon vanilla extract
1 large egg, beaten
2¼ cups sifted flour
2 teaspoons baking powder
¾ cup chopped nuts

Preheat oven to 350°.

9 x 5 loaf pan

Grease and flour loaf pan. Place apricots, raisins, and grated rind in large mixing bowl. Add boiling water to orange juice to make 1 cup liquid; add to fruit. Stir in baking soda, sugar, butter, and vanilla. Stir in egg. Sift in flour and baking powder; stir. Blend in nuts. Pour into prepared pan and bake for 50 minutes or until lightly browned. Turn out on wire rack to cool.

BLUEBERRY LEMON BREAD

Yield: 1 loaf

1½ cups all-purpose flour
1 teaspoon baking powder
¼ teaspoon salt
6 tablespoons unsalted butter (room temperature)
1⅓ cups sugar, divided
2 large eggs

2 teaspoons grated lemon peel
½ cup milk
1½ cups fresh blueberries or frozen blueberries, thawed and well drained
3 tablespoons fresh lemon juice

Preheat oven to 325°.

9 x 5 loaf pan

Grease loaf pan. Combine flour, baking powder, and salt in small bowl. In a large bowl, using electric mixer, cream butter with 1 cup sugar until mixture is light and fluffy. Add eggs one at a time, beating well after each addition. Add lemon peel. Mix in dry ingredients alternately with milk, beginning and ending with dry ingredients. Fold in blueberries. Spoon batter into prepared pan. Bake until golden brown and toothpick inserted in center comes out clean, about 1 hour and 15 minutes. During last 15 minutes of baking time, bring ⅓ cup sugar and 3 tablespoons lemon juice to a boil in a small saucepan, stirring until sugar dissolves. Pierce top of hot loaf several times with a toothpick. Pour hot lemon mixture over loaf while still in pan. Cool 30 minutes in pan on rack. Turn bread out of pan; cool completely on rack.

BANANA BREAD

Yield: 2 loaves

2 cups flour
¼ teaspoon salt
1 teaspoon baking soda
1½ cups sugar

1 cup margarine
2 eggs, beaten
3 bananas, mashed
1 cup chopped nuts (optional)

Preheat oven to 325°.   2 (8 inch) loaf pans

Grease and flour loaf pans. Sift flour, salt, and soda together; set aside. Cream sugar and margarine. Add eggs and mashed bananas. Add dry ingredients and nuts; mix well. Pour batter in loaf pans. Bake for 55 minutes, or until toothpick comes out clean.

BUTTERMILK BISCUITS

Yield: 3-4 dozen

1 package dry yeast
1 cup lukewarm water
7 cups flour
2 cups buttermilk

4 teaspoons baking powder
½ cup sugar
¼ teaspoon soda
½ cup oil

Preheat oven to 400°.   Cookie sheet

Dissolve yeast in lukewarm water. Combine flour, buttermilk, and yeast mixture in a large mixing bowl. Add remaining ingredients. Dough may be rolled out and cut with cutter, or biscuits may be dropped by heaping tablespoon onto the cookie sheet. Bake for 25-30 minutes or until brown.

PUMPKIN BREAD

Serves: 24

⅔ cup water
1 cup vegetable oil
4 eggs
1 (16 oz.) can pumpkin
3 cups sugar
3⅓ cups flour

2 teaspoons baking soda
1½ teaspoons salt
1 teaspoon cinnamon
1 teaspoon nutmeg
1 teaspoon vanilla extract

Preheat oven to 350°.   3 (9 x 5) loaf pans

Grease and flour loaf pans. Mix water, oil, eggs, pumpkin, and sugar. Blend well; add remaining ingredients. Pour into prepared pans. Bake for 1 hour. Remove from pans to cool.

GINGERBREAD

Serves: 8

1 cup molasses	1 teaspoon ginger
1½ cups boiling water	2 teaspoons cinnamon
1 teaspoon baking soda	2½ cups flour, sifted
½ cup butter	1 tablespoon baking powder
1 cup sugar	1 egg, well beaten
½ teaspoon salt	

Preheat oven to 375°.   2 (11 x 7) pans

Grease pans. Combine the molasses with the boiling water and baking soda. Allow to cool. Cream butter with sugar. Add molasses water to the creamed butter. Add the salt, ginger, and cinnamon; beat well. Sift flour with baking powder. Add egg and flour to the mixture. Batter should be very thin. If the batter is not very thin, add a little milk so that it runs off the spoon. Pour into prepared pans; bake for 20 minutes. Reduce heat to 300° and bake for 20 minutes more.

Milk may be substituted for water.

HUSHPUPPIES

Serves: 10-12

Vegetable oil sufficient for frying	2 tablespoons chopped green chilies
2 cups cornmeal	¼ medium green bell pepper, chopped
1 cup flour	1 small onion, chopped
3 eggs	Pinch of baking soda
3 teaspoons baking powder	⅛ cup (more or less) buttermilk
1½ teaspoons salt	
2 (16½ oz.) cans cream style corn	

Heat oil for frying.   Deep fat fryer or deep skillet

Mix all ingredients except buttermilk. Add buttermilk slowly, and mix batter to a consistency of cornbread batter. Drop into medium hot oil and fry until golden brown. Hints for success: If the batter is too heavy and the hushpuppies do not rise to the top of the oil, add more baking powder. If the batter is too greasy and they fall apart while frying, add flour. Serve hot.

FIESTA GRAPEFRUIT BREAD

Yield: 2 loaves

1 Texas red grapefruit
⅔ cup shortening
1 cup brown sugar, firmly packed
1½ cups sugar
1 teaspoon vanilla extract
4 eggs
1 (16 oz.) can pumpkin

3½ cups flour
½ teaspoon baking powder
2 teaspoons baking soda
2 teaspoons salt
2 teaspoons cinnamon
½ teaspoon cloves
½ teaspoon nutmeg
1 cup chopped pecans

Preheat oven to 350°. 2 (9 x 5) loaf pans

Grease loaf pans. Coarsely grate all outer peel from the grapefruit; set aside. Remove remaining white peel and seeds; puree whole grapefruit in blender or food processor. Do not drain, set grapefruit aside with peel.

In a large bowl, cream shortening, sugars, and vanilla well. Mix in eggs and pumpkin. Combine flour, baking powder, soda, salt, and spices; gradually stir into pumpkin mixture. Fold in grapefruit puree, the peel, and pecans. Pour into pans. Bake for about 70 minutes, or until done. Turn out on wire rack to cool thoroughly.

This recipe can easily be cut in half to make one loaf or 18 muffins.

CRANBERRY ORANGE BREAD

Yield: 1 loaf

2 cups sifted flour
¾ cup sugar
1½ teaspoon baking powder
1 teaspoon salt
½ teaspoon soda
1 cup cranberries, coarsely cut

½ cup pecans, chopped
1 teaspoon orange peel, grated
1 egg, beaten
¾ cup orange juice
2 tablespoons salad oil

Preheat oven to 350°. 9 x 5 loaf pan

Grease loaf pan. In a large bowl, sift together first 5 ingredients. Stir in cranberries, pecans and orange peel. Combine egg, juice and oil in a small bowl. Add to dry ingredients, stirring just until moistened. Bake for 50 minutes. Remove from pan and cool. This is great sliced and spread with cream cheese or butter.

LEMON CRANBERRY LOAF

Yield: 1 loaf

1 (8 oz.) package cream
cheese, softened
⅓ cup margarine
1¼ cups sugar
1 teaspoon vanilla extract
3 eggs

2 tablespoons lemon juice
1½ cups chopped cranberries
2¼ cups flour
2 teaspoons baking powder
½ teaspoon baking soda

Preheat oven to 325°.

9 x 5 loaf pan

Grease and flour loaf pan. Combine cream cheese, margarine, sugar, and vanilla in a large mixing bowl; beat well with an electric mixer. Add eggs, one at a time, mixing after each addition. Stir in lemon juice. Toss cranberries, flour, baking powder, and baking soda together in a large bowl. Add cream cheese mixture and mix just until moistened. Pour mixture into prepared pan. Bake for 1 hour and 15 minutes.

Glaze:

1½ cups sifted powdered sugar
4 tablespoons lemon juice

2 tablespoons butter, softened

Combine powdered sugar, lemon juice, and butter to make glaze. Allow bread to cool 5 minutes; glaze. Remove from pan; cool completely.

TEXAS-SIZE BISCUITS

Yield: 12-14 biscuits

3 cups flour
2 tablespoons sugar
4½ teaspoons baking powder
¾ teaspoon cream of tartar

¾ teaspoon salt
¾ cup shortening
¾ cup milk
1 egg, beaten

Preheat oven to 450°.

Cookie sheet

Combine flour, sugar, baking powder, cream of tartar, and salt; cut in shortening. Add milk and egg; stir until moist. Knead 8 or 10 times on lightly floured surface. Roll dough to ¾ inch thickness. Cut with biscuit cutter. Place on ungreased cookie sheet and bake 12-15 minutes.

ORANGE BREAD

Yield: 1 loaf

1 large orange
Boiling water, up to 1 cup
Raisins or dates, up to 1 cup
2 tablespoons butter, melted
1 cup sugar
1 egg, beaten

1 teaspoon vanilla extract
2 cups flour
¼ teaspoon salt
1 teaspoon baking powder
1 teaspoon baking soda
¾ cup chopped pecans

Preheat oven to 350°.   Loaf pan

Lightly grease loaf pan. Remove orange peel from orange with a sharp knife, scraping away all of the white pectin from the peel. Cut orange in half and squeeze juice into measuring cup; add boiling water to make 1 cup. Grind orange peel in food processor. Grind dates or raisins in food processor and add to orange peel to make 1 cup. Cream butter and sugar; add egg and vanilla. Sift dry ingredients together, and add to creamed mixture alternating with peel mixture and juice. Beat well; add nuts. Bake for 50 minutes. Allow to cool in pan.

POPPY SEED BREAD

Serves: 14

3 cups flour
1½ teaspoons salt
1½ teaspoons baking powder
2½ cups sugar
1½ teaspoons vanilla extract

1½ teaspoons almond extract
1½ cups milk
1½ cups vegetable oil
1½ tablespoons poppy seeds
3 eggs

Preheat oven to 350°.   2 (9 x 5) loaf pans

Grease and flour loaf pans. Mix the bread ingredients in a large bowl and beat thoroughly. Pour into prepared loaf pans. Bake for 1 hour and 15 minutes.

Glaze:
½ cup sugar
¼ cup orange juice
½ teaspoon vanilla extract

½ teaspoon almond extract
½ teaspoon butter flavoring

Mix the glaze ingredients together and pour over hot bread. Easy!

SOUR CREAM COFFEE BREAD

Serves: 12-16

1 cup margarine
2 cups sugar
2 eggs
1 cup sour cream
½ teaspoon vanilla extract
2 cups flour

1 teaspoon baking powder
¼ teaspoon salt
4 teaspoons sugar
1 teaspoon cinnamon
1 cup chopped pecans

Preheat oven to 350°.   9 x 3 tube pan

Cream margarine and sugar; beat in eggs one at a time. Mix thoroughly. Beat in sour cream and vanilla. Sift flour, baking powder, and salt together; fold into mixture. Combine sugar with cinnamon and nuts. Spoon about ⅓ of the batter into pan. Sprinkle with about ¾ of the pecan mixture. Spoon in remaining batter; sprinkle on remaining pecan mixture. Bake for 1 hour, or until done. Remove from pan. Cool on wire rack.

STRAWBERRY BREAD

Serves: 14

3 cups all-purpose flour
2 cups sugar
1 tablespoon cinnamon
1 teaspoon baking soda
1 teaspoon salt

3 eggs, well beaten
1¼ cups vegetable oil
2 (10 oz.) packages frozen
   sliced strawberries, thawed

Preheat oven to 350°.   2 (9 x 5) loaf pans

Lightly grease loaf pans and line the bottom of each pan with foil. In a large bowl, mix together flour, sugar, cinnamon, baking soda, and salt. Make a well in the center of this mixture. Pour eggs and oil in the well. Stir until dry ingredients become moist. Pour thawed strawberries and juice into a separate container. With a slotted spoon, dip out strawberries and gently stir into the above mixture; add juice gradually. Stir until batter is of cake consistency. The amount of juice varies in frozen strawberries and you do not want the batter to be too thin. Bake for 1 hour or until tests done. Let cool for 15 minutes. With a knife loosen sides of loaves from pans. Turn out of pans onto rack. Let loaves completely cool before slicing.

 Cut bread with a hot knife; it will keep its shape.

TROPICAL BREAD

Serves: 9

⅔ cup sugar
⅓ cup shortening, room
   temperature
2 eggs, room temperature
1 cup ripe banana
¼ cup buttermilk
1¼ cups all-purpose flour

1 teaspoon baking powder
½ teaspoon salt
½ teaspoon baking soda
1 cup 100% bran cereal
¾ cup dried apricots
½ cup chopped walnuts

Preheat oven to 350°.　　　　　　　　　　　　　9 x 5 loaf pan

Grease loaf pan and line with buttered wax paper to prevent the loaf from sticking. In a large bowl, cream sugar and shortening, and add eggs. Beat thoroughly. In a small bowl, mash banana well and combine with buttermilk. Sift together the flour, baking powder, salt, and baking soda; add alternately with banana mixture to the creamed mixture. Stir in the bran, apricots, and walnuts. Stir only until well blended. Pour into prepared pan. Bake for 1 hour or until it tests done when pierced with a toothpick. It will probably have a crack running the length of the crust. Remove bread from the oven. Let loaf cool in the pan for about 10 minutes. Turn out on wire rack. This bread is even better the second and third day.

ZUCCHINI NUT BREAD

Serves: 8-10

3 eggs
1 cup vegetable oil
2½ cups sugar
2 cups peeled and grated
   zucchini
3 tablespoons vanilla extract

1 teaspoon baking soda
1 teaspoon baking powder
1 teaspoon nutmeg
1 tablespoon cinnamon
3 cups flour
1 cup chopped nuts

Preheat oven to 325°.　　　　　　　　　　2 (9 x 5) loaf pans

Grease loaf pans. Beat eggs; add oil, sugar, zucchini, and vanilla. Mix well. Add remaining ingredients; mix well. Pour into prepared loaf pans. Bake for 1 hour. Great for breakfast or finger sandwiches with cream cheese.

 Freeze over-ripe bananas in the peel for use in bread and muffin recipes.

APPLESAUCE PUFFS
Yield: 8-10 muffins

2 cups Bisquick
¼ cup sugar
1 teaspoon cinnamon
½ cup applesauce

¼ cup milk
1 egg
2 tablespoons oil

Preheat oven to 400°.　　　　　　　　　　　　　Muffin tins

Grease muffin tins. Mix together dry ingredients. Add applesauce, milk, egg, and oil. Beat well for 1 minute. Fill muffin cups ⅔ full; bake for 12 minutes. If desired, while hot, dip tops in melted butter, then in a cinnamon-sugar mixture.

CARROT-ORANGE MUFFINS
Yield: 12 regular or 36 miniature muffins

1 cup unbleached all-purpose
　flour
1 cup whole-wheat flour
2 teaspoons baking powder
1 teaspoon ground cinnamon
¼ teaspoon salt
1 teaspoon freshly grated
　orange peel

½ cup plus 2 tablespoons
　orange juice, preferably
　freshly squeezed
½ cup skim milk
¼ cup vegetable oil
2 tablespoons honey
　Whites of 2 large eggs
1 cup coarsely grated carrots

Preheat oven to 400°.　　　　Regular or miniature muffin tins

Grease muffin tins or line with paper baking cups. Thoroughly mix flours, baking powder, cinnamon, and salt in a large bowl. Put orange peel, orange juice, milk, oil, honey, and egg whites in a medium size bowl. Beat with a fork or whisk until egg whites are broken up (the mixture will have a curdled look). Stir in the carrots. Pour over dry ingredients and fold in with a rubber spatula just until dry ingredients are moistened. Scoop batter into prepared muffin cups. Bake 20-25 minutes, or until lightly browned. Place pan on a wire rack to cool for a few minutes before removing muffins.

Variation: For zucchini-orange muffins, use 1 cup coarsely grated zucchini instead of the carrot.

CRANBERRY-PECAN MUFFINS

Serves: 24

Muffins:
½ cup butter or margarine	½ teaspoon nutmeg
1 cup sugar	1 (8 oz.) carton sour cream
2 eggs	1 teaspoon vanilla extract
2 cups sifted flour	¾ cup chopped cranberries
1 teaspoon baking powder	⅓ cup chopped nuts
½ teaspoon baking soda	

Topping:
2 tablespoons sugar	¼ teaspoon nutmeg

Preheat oven to 350°.   Muffin tin

Grease or line muffin pan with paper cups. Cream butter and sugar well. Add eggs one at a time. Sift together flour, baking powder, soda, and nutmeg. Add to creamed mixture alternately with sour cream. Stir in vanilla. Fold in cranberries and nuts. Sprinkle with topping before baking. Bake about 20 minutes.

CREAM OF WHEAT MUFFINS

Yield: 12 muffins

1 cup of flour, sifted	1 teaspoon salt
1 cup regular cream of wheat cereal	3 teaspoons baking powder
	1 egg
3 tablespoons sugar, rounded slightly	1 cup milk
	⅓ cup oil

Preheat oven to 375°.   Muffin pan

Grease or line muffin pan with paper cups. Combine flour, cream of wheat, sugar, salt, baking power, egg, milk, and oil. Put in muffin tins; bake for 30 minutes.

 Muffins will slide out of tin pans if the hot pan is first placed on a wet towel.

LEMON SLICE MUFFINS

Yield: 12 muffins

3 lemons
1 tablespoon water
½ cup sugar
6 tablespoons unsalted butter
2 cups all-purpose flour

2 teaspoons baking powder
½ teaspoon baking soda
½ teaspoon salt
2 eggs
1 cup milk

Preheat oven to 400°.

Muffin tin

Grease muffin tins. Finely grate the yellow part of the peel from the lemons. Set lemons aside. Combine peel, water and ¼ cup sugar in small saucepan. Stir over medium heat for 2 minutes, until sugar dissolves. Add butter and stir until melted. Set aside. Discard remaining peel from lemons. Cut lemons crosswise into slices about ¼ inch thick; discard seeds. You will need 12 lemon slices. Put 1 teaspoon of remaining sugar in bottom of each muffin cup and place lemon slice on top.

In medium bowl, mix flour, baking powder, baking soda and salt. Set aside. In larger bowl, whisk eggs, milk and reserved lemon mixture until smooth. Add dry ingredients and mix until blended. Spoon into tins and bake 15-20 minutes until toothpick comes out clean. Invert tin onto wire rack and cool for 5 minutes before lifting away tin.

STRAWBERRY-ORANGE MUFFINS

Yield: 12 muffins

2¼ cups all-purpose flour
2 teaspoons baking powder
1 teaspoon baking soda
½ teaspoon salt
¾ cup sugar
½ cup milk
½ cup sour cream

⅓ cup vegetable oil
1 egg
1 tablespoon finely grated orange zest
1 cup thinly sliced fresh strawberries
⅓ cup strawberry jam

Preheat oven to 400°.

Muffin tin

Grease muffin tins. In large bowl, mix flour, baking powder, baking soda, and salt. Set aside. In medium bowl, whisk together sugar, milk, sour cream, oil, egg and orange zest. Pat strawberry slices dry between paper towels and stir into sugar mixture. Add to dry ingredients and stir until blended. Place a spoonful of batter into each cup. Top with a scant teaspoon of strawberry jam. Fill tins two-thirds full with remaining batter. Bake 15-18 minutes until toothpick comes out clean. Cool in tins 5 minutes before removing.

RASPBERRY STREUSEL MUFFINS

Yield: 12 muffins

Batter:

1½ cups all-purpose flour
¼ cup granulated sugar
¼ cup dark brown sugar, firmly packed
2 teaspoons baking powder
¼ teaspoon salt
1 teaspoon ground cinnamon

1 egg, lightly beaten
½ cup unsalted butter, melted
½ cup milk
1¼ cups fresh raspberries or 1 bag frozen raspberries, drained
1 teaspoon grated lemon zest

Preheat oven to 350°.   Muffin pan

Line muffin pan with paper baking liners. Sift flour, sugar, brown sugar, baking powder, salt, and cinnamon together in a medium size bowl. Make a well in the center of the mixture. Place the egg, melted butter, and milk in the well. Stir with a spoon just until ingredients are combined. Stir in raspberries and lemon zest. Do not over mix. Fill each muffin cup ¾ full of batter.

Streusel Topping:

½ cup chopped walnuts or pecans
½ cup dark brown sugar, firmly packed
¼ cup all-purpose flour

1 teaspoon ground cinnamon
1 teaspoon grated lemon zest
2 tablespoons unsalted butter, melted

Combine nuts, brown sugar, flour, cinnamon, and lemon zest in a small bowl. Pour in melted butter; stir to combine. (You may want to reserve half of this topping for a later date.)
Sprinkle topping over top of each muffin. Bake 20-30 minutes, until golden brown and firm.

Glaze:

½ cup powdered sugar   1 tablespoon fresh lemon juice

Mix sugar and lemon juice; drizzle over warm muffins.

COUNTRY APPLE COFFEE CAKE

2 tablespoons butter, divided
1⅓ cups peeled and chopped
apples, divided
1 (10 oz.) can Hungry Jack
Flaky Biscuits
⅓ cup brown sugar, firmly
packed

¼ teaspoon cinnamon
⅓ cup light corn syrup
1⅓ teaspoons whiskey (optional)
1 egg
½ cup pecan pieces

Preheat oven to 350°.   9 inch round or 8 inch square pan

Grease bottom and sides of pan with 1 tablespoon of butter. Spread one cup of apples in pan. Cut each biscuit into four pieces. Arrange point side up over apples. Top with remaining apples. Mix remaining butter with brown sugar, cinnamon, corn syrup, whiskey, and egg for 2 minutes. Stir in pecans. Spoon over biscuits. Bake 35-45 minutes until deep golden brown. Cool for 5 minutes, then drizzle glaze over cake.

Glaze:

⅓ cup powdered sugar
¼ teaspoon vanilla extract

2 teaspoons milk

Mix powdered sugar, vanilla, and milk together until smooth.

SPANISH COFFEE CAKE

Serves: 12

2½ cups flour
1 cup brown sugar, firmly
packed
¾ cup sugar
1 teaspoon cinnamon
½ teaspoon salt

¾ cup vegetable oil
⅓ cup chopped nuts
1 teaspoon baking soda
1 teaspoon baking powder
1 cup buttermilk
1 egg

Preheat oven to 350°.   13 x 9 baking pan

Mix flour, sugars, cinnamon, salt, and oil to make a crumbly mixture. Remove ½ cup; mix with nuts to use as the topping. Mix the soda, baking powder, buttermilk, and egg with the remaining crumbly mixture. Pour into pan; sprinkle with the topping. Bake for 30 minutes. Test by inserting toothpick before removing from oven.

CRANBERRY CAKE

Serves: 12

½ cup margarine
1 cup sugar
1 egg or egg substitute
2 egg whites
½ teaspoon almond extract

2 cups flour
1 teaspoon baking powder
1 cup nonfat plain yogurt
1 cup whole cranberry sauce
½ cup sliced almonds

Preheat oven to 350°.   13 x 9 pan

Lightly grease and flour pan. Cream margarine; gradually add sugar, beating until fluffy. Add egg and egg whites, beating after each. Add almond extract. Combine flour and baking powder. Add flour mixture and yogurt, alternately, to sugar mixture, beginning and ending with flour mixture. Pour batter in pan. Spoon cranberry sauce evenly over batter; sprinkle with almonds. Bake 25 minutes, or until cake pulls slightly away from sides of pan.

Glaze:

1 cup powdered sugar
2 tablespoons skim milk

½ teaspoon vanilla extract

Combine above ingredients; drizzle over warm cake.

Freshen dry, crusty rolls or French bread by sprinkling with a few drops of water, wrap in aluminum foil, and reheat at 350° for about 10 minutes.

RUBY SWEET COFFEE CAKE

Serves: 10-12

2 Texas red grapefruits
1 cup quick cooking oats
½ cup margarine
1 cup sugar
½ cup brown sugar, firmly
   packed

2 eggs, beaten
1⅓ cups flour
1 teaspoon baking soda
½ teaspoon cinnamon
   Vanilla yogurt or whipped
   cream, optional

Preheat oven to 350°.   13 x 9 baking pan

Grease baking pan. Finely grate 2 teaspoons peel from grapefruit and set aside. Peel and section grapefruit over bowl, reserving juice. Set sections aside. Measure grapefruit juice, and add water to equal 1¼ cups liquid. Bring liquid to a boil in a saucepan. Remove from heat and stir in oats and butter; let stand 20 minutes. Combine reserved grapefruit peel and remaining ingredients, except yogurt, in a mixing bowl. Stir in oats mixture, blending well. Pour into baking pan. Bake for 35 minutes. Serve warm or cool, topped with yogurt or whipped cream. Garnish with grapefruit sections.

This recipe makes great muffins, too!

Parsley, chopped finely, is a wonderful addition to biscuits, the crust of a chicken pot pie, or quiche.

Main Entrees

Refer to Mexican Cuisine section for additional recipes in this category.

 *This recipe is pictured on the previous page.*

GRILLED BRISKET

Serves: 10-12

6-8 pound brisket of beef

Marinade:

1 (32 oz.) bottle ketchup	Juice of ½ lemon
5 tablespoons Worcestershire sauce	½ cup butter
1 tablespoon Tabasco sauce	1 small onion, finely chopped
Salt and lemon pepper to taste	

Prepare grill. Mix all ingredients for marinade. Baste brisket with marinade, grilling over low flames. Use marinade to keep meat moist throughout grilling process, approximately 30-45 minutes, until brisket is browned. Wrap brisket in foil and continue basting and cooking slowly over low heat (1 hour per pound).

Sauce:

1 cup vinegar	2 tablespoons Worcestershire sauce
½ cup butter	Juice of 1 lemon
Salt and lemon pepper to taste	1½ teaspoons Tabasco sauce
1 clove garlic, minced	

When brisket is done, slice meat and pour sauce mixture over meat. Place in pan; cover well. Heat in low temperature oven (275° to 300°) for one hour.

 The key to the best flavor in grilling is a powerful, even source of heat. Cooking should not begin until the coals have been burning long enough to be covered with gray ash. Allow at least 35-45 minutes after lighting the coals for the ash coating to develop.
The grill must be absolutely clean and free of residue from previous barbecues!

SMOKED BRISKET

Serves: 8-10

6 pound brisket of beef
3 tablespoons Liquid Smoke
3 teaspoons garlic salt or
  powder
3 teaspoons onion salt

3 teaspoons celery salt
3 tablespoons Worcestershire
  sauce
Pepper to taste

Cover brisket with the first four ingredients. Let this stand refrigerated overnight in an oven baking bag. The next morning cover with the remaining two ingredients. Bake at 275° for 5 hours.

Barbeque Sauce:
  1 cup ketchup
  1 teaspoon salt
  1 teaspoon celery seed
  ¼ cup brown sugar

  ¼ cup Worcestershire sauce
  2 cups water
  1 onion, chopped
  ¼ cup vinegar

Mix all ingredients together and boil for 15 minutes. Pour over brisket and bake 1 hour longer.

NEVER FAIL MEATLOAF

Serves: 6-8

1 pound ground beef
1 teaspoon salt
1 tablespoon brown sugar
1 tablespoon Worcestershire
  sauce
¼ teaspoon paprika
½ cup ketchup
½ cup milk

1 cup cracker crumbs
1 egg
1 small onion, diced
½ medium green bell pepper,
  chopped
1 (10¾ oz.) can cream of
  mushroom soup

Preheat oven to 350°.   9 x 5 loaf pan

Lightly grease loaf pan. Mix all ingredients (except soup) together. Pat into pan. Top with soup. Bake for 1-1½ hours.

 Meatloaf will not stick if you place a slice of bacon on the bottom of the pan.

BEEF AND BLACK BEAN CHILI

Serves: 6

1⅓ cups black beans
1¼ teaspoons cumin
1¼ teaspoons dried oregano
1 small bay leaf
1 tablespoon olive oil
1 medium onion, chopped
1 medium green bell pepper, coarsely chopped
1 medium red bell pepper, coarsely chopped
1½ tablespoons jalapeño pepper, finely chopped
1 pound boneless top sirloin or top round, cut in one inch cubes

2 (14½ oz.) cans tomatoes, undrained and chopped
1½ tablespoons red wine vinegar
⅓ teaspoon black pepper
1 tablespoon finely chopped parsley
⅓ teaspoon cayenne pepper
⅔ teaspoon paprika
4 cups black bean cooking liquid
1 clove garlic, minced

Place black beans in heavy pot; cover with cold water at least 2 inches above beans. Bring to a boil; simmer until tender (approximately 1¼ hours). Add water if necessary. When tender, drain beans and reserve liquid. Combine cumin, oregano, and bay leaf. Heat oil in skillet and sauté onions, green, red, and jalapeño peppers. Sauté 1 to 2 minutes. Remove vegetables to heavy pot. In skillet, add beef; brown. Add to vegetables. Add all remaining ingredients to beef/vegetable mixture. Bring to a boil; simmer until beef is tender, approximately 1 hour.

ITALIAN BEEF BAKE

Serves: 4

2 pounds sirloin, cut into 2-inch strips
1 envelope (1⅜ oz.) onion soup mix
1 green bell pepper, chopped
¼ cup chopped onion
1 (16 oz.) can tomatoes

Dash Pepper
¼ teaspoon salt
1 tablespoon bottled steak sauce
1 tablespoon cornstarch
2 tablespoons chopped parsley

Preheat oven to 350°.   11 x 9 baking pan

Line baking pan with enough foil to cover meat. Place meat on foil. Mix remaining ingredients in large mixing bowl and pour over meat. Seal foil and bake for 1½-2 hours.

CORNBREAD PIE

Serves: 6

1 cup cornmeal
2 eggs
1 (14 oz.) can sweetened condensed milk
½ teaspoon salt
1 (16½ oz.) can creamed corn
¼ cup bacon drippings (or bacon bits)

¼ teaspoon baking soda
1½ pounds ground beef
1 large onion, finely chopped
2 tablespoons additional bacon drippings
4 jalapeños, seeded; chopped
½ pound Monterey Jack cheese, grated

Preheat oven to 450°.   11 x 7 baking pan

Grease baking pan. Mix first 7 ingredients in large mixing bowl. Brown ground meat; drain. Add 2 tablespoons drippings and onion to meat and cook until onion is transparent. Add jalapeños to beef mixture. Pour half of cornmeal mixture into pan. Spread all of ground beef mixture over this and top with cheese. Pour remaining cornmeal mixture over cheese. Bake for 45 minutes. Spread butter over top immediately after removing from oven, if desired.

SESAME STEAK

Serves: 4-6

Marinade:

⅓ cup sesame seeds
½ cup vegetable oil
4 medium onions, thickly sliced
½ cup soy sauce

¼ cup lemon juice
1 tablespoon sugar
½ teaspoon pepper
2 cloves garlic, minced

Brown sesame seeds and combine with remaining ingredients. Reserve ⅓ cup to serve with steak, if desired. Place in a large resealable plastic bag.

Meat:
2½-3 pound sirloin steak-1½ inch thick

Preheat broiler   Broiler pan

Trim fat from steak. Place steak in plastic bag of marinade turning to coat both sides. Marinate in refrigerator at least five hours. Place meat on broiler pan and cook to desired doneness. Heat reserved marinade and serve with steak.

 To quickly remove food stuck to a casserole dish, fill with boiling water and 2 tablespoons baking soda or salt.

PEPPER STEAK

Serves: 4

2 pounds lean beef tenderloin
1 tablespoon cracked pepper
3 tablespoons butter
½ teaspoon olive oil
1 cup sour cream

½ cup red wine
1 tablespoon powdered beef stock
1 tablespoon Kitchen Bouquet

Slice beef into 4 fillets; dredge meat in pepper. Refrigerate for 6 hours. Preheat oven to 350°. Melt butter and olive oil in heavy skillet. Brown meat for 2 minutes on each side; remove from skillet; place in casserole. Bake for 10 minutes. Remove from oven; cover. In the same skillet, add sour cream, wine, beef stock, Kitchen Bouquet, and any remaining pepper. Cook sauce for about 15 minutes until thickened. Place meat on individual plates; spoon sauce on top and serve.

BEEFY ONION ROAST

Serves: 8

4-4½ pound rump roast
Vegetable oil
Flour and pepper to coat roast
1 envelope Lipton Beefy Onion Soup
½ onion, sliced in rings

1 small carton fresh mushrooms or 1 (4 oz.) can drained mushrooms
¼-½ teaspoon coarse pepper
1 tablespoon Wonder flour (optional)

Pour enough oil to cover the base of a large pot or pan; heat. Wash roast; coat in flour and pepper mixture. Brown roast on all sides on medium high heat. While browning, mix soup with 1½ cups water. Lower heat; add soup mix, onions, mushrooms, and pepper. Cook on low heat until tender, approximately 2-3 hours. Turn 1 to 3 times during cooking. It will make its own gravy. When ready to serve, remove roast; slice. If thicker gravy is preferred, put Wonder Flour in a mixing cup, add a little gravy at a time to mix with flour. Add the mixing cup gravy to pot gravy; stir.

 A little salt sprinkled into the frying pan will prevent spattering.

SPICY STEAK

Serves: 6

1½ pounds lean top sirloin steak
Seasoned salt, pepper, and garlic powder
1 medium onion, chopped
1 small green bell pepper, chopped
1 garlic clove, minced

1 (8 oz.) can tomato sauce
¼ cup Burgundy wine
1 tablespoon minced parsley
½ teaspoon dried Italian seasoning
1 (14½ oz.) can Mexican-style tomatoes

Trim fat from steak; cut into 6 equal pieces. Sprinkle steak with seasoned salt, pepper, and garlic powder. In a large skillet, which has been coated with non-stick spray, brown steaks on both sides on medium high heat. Remove steaks from skillet; re-spray if needed. Sauté onion, green pepper, and garlic until soft. Add tomato sauce, wine, parsley, Italian seasoning, and tomatoes. Mix well. Return steaks to skillet. Cover, reduce heat, and simmer 1½ hours, or until tender.

Lean round steak may be used, but is not as tender.

STEAK DIANE

Serves: 4

4 (½ inch thick) rib eye beef steaks
Salt and pepper to taste
4 tablespoons butter or margarine, divided
¼ cup brandy, divided
2 small shallots, minced and divided

3 tablespoons chopped chives, divided
½ cup dry sherry, divided
Cherry tomatoes and celery leaves to garnish, divided

About 20 minutes before serving, pound steaks to ¼ inch thick with meat mallet or edge of plate on cutting board, turning occasionally. Season meat on both sides with salt and pepper. Cook 1 steak at a time in skillet in 1 tablespoon hot butter or margarine until both sides are brown. Pour 1 tablespoon brandy over steak; set aflame with match. When flaming stops, stir in ¼ each of shallots and chives; stir constantly until shallots and chives are tender. Add 2 tablespoons sherry; heat through. Place steak on warm plate; pour sherry mixture over top. Garnish with split cherry tomatoes and celery leaves. Serve immediately; repeat process with remaining steaks.

STEAK WITH MUSTARD SAUCE

Serves: 4

1 tablespoon butter
1 tablespoon vegetable oil
4 (¾ inch thick) rib steaks or
  filets
  Salt and freshly ground
  pepper

2 shallots, minced
2 tablespoons Dijon mustard,
  or to taste
½ pint whipping cream

Melt butter and vegetable oil in large heavy skillet over high heat. Season steaks on all sides with salt and pepper. Place steaks in skillet; cook about 3 minutes on each side; transfer steaks to plates, tent with foil. Pour off all but film of fat from skillet; add shallots, sauté over medium heat 2 minutes. Whisk in mustard, then cream. Bring to a boil; spoon over steaks.

TENDERLOIN WITH PEPPERCORN SAUCE

Serves: 4

4 pounds beef tenderloin
2 tablespoons vegetable oil
  Salt and black pepper to
  season
3 tablespoons heated cognac

1½ cups brown stock/beef broth
½ pint whipping cream
3 tablespoons soaked green
  peppercorns
3 tablespoons butter

Preheat oven to 450°.   Small roasting pan

Brown the beef tenderloin on all sides in vegetable oil over medium heat in a large heavy skillet. Season to taste with salt and pepper. Transfer beef to a roasting pan; bake covered for 40 minutes (10 minutes per pound for rare), or longer if desired. Put beef on a cutting board; let stand for 5-10 minutes. Pour off fat from skillet, return to heat and add heated cognac. Ignite (optional); shake skillet until flame goes out. Stir in brown stock and whipping cream. Reduce sauce to desired thickness; add peppercorns and butter. Serve meat in ½ inch slices with sauce.

 When browning any piece of meat, the job will be done more quickly and effectively if the meat is very dry and the fat is very hot.

LOBSTER-STUFFED TENDERLOIN

Serves: 10-12

3 (4 oz.) lobster tails, frozen
1 (4-6 lbs.) beef tenderloin
1 tablespoon butter or
  margarine, melted

1½ teaspoons lemon juice
Garlic salt
Freshly ground pepper
6 slices bacon

Preheat oven to 425°.   Shallow roasting pan

In a large pot bring salted water to boil. Once water is boiling, drop lobster tails in and return to boil; simmer for 5 minutes. Carefully remove lobster from tail in one piece and set aside. Cut tenderloin lengthwise to within ½ inch of edge, leaving one side connected. Place lobster inside tenderloin to cover entirely. Combine butter and lemon juice and drizzle over lobster. Wrap top of tenderloin around the lobster and tie with string. Sprinkle with garlic salt and pepper. Place on rack in roasting pan. Bake uncovered for 35 to 40 minutes or until meat thermometer reads 140° to 160°. Meanwhile cook bacon until transparent, not crisp. Set aside to cool. Place bacon on top of tenderloin for the last 4-5 minutes. May be served with this optional sauce.

Sauce:
½ cup sliced green onions
½ cup butter or margarine,
  melted

½ cup dry white wine
⅛ teaspoon garlic salt

In a small saucepan, combine onions, butter, wine, and garlic salt; cook over low heat. Strain and serve on top of beef or on the side as a sauce.

SPECIAL BAKED TENDERLOIN

Serves: 8-10

1 (4 lbs.) large beef tenderloin
Margarine
1 beef bouillon cube

1 cup red wine
1 pound fresh mushrooms,
  whole or sliced

Preheat oven to 400°.   Roasting pan

Place tenderloin in roasting pan with at least 2 inch sides. Cover top and sides of tenderloin with margarine. Bake uncovered for 30 minutes. Dissolve beef bouillon in red wine and pour over tenderloin; add mushrooms and cook for 20 more minutes (no longer). You will have some rare, medium rare, and medium meat.

STUFFED TENDERLOIN

Serves: 6-8

½ cup butter
¾ cup finely chopped
  mushrooms
1 (4-6 oz.) package ham, diced
½ cup minced green onion

¼ teaspoon salt
¼ teaspoon pepper
3 cups white bread cubes
2 tablespoons water
1 (2½ lb.) tenderloin, slit

Preheat broiler.

Melt butter over medium heat; add mushrooms, ham, onions, salt, and pepper. Cook until tender. Add bread crumbs and water; toss gently. Fill slit in tenderloin with this mixture. Tie; broil 15 minutes on each side.

Sauce:

1 (10 oz.) can tomatoes
4 tablespoons butter
2 tablespoons flour
1 (13¾ oz.) can beef broth
1 tablespoon brandy

½ teaspoon pepper
½ teaspoon meat extract paste,
  optional
¼ teaspoon salt

Drain, reserving liquid, and finely chop tomatoes. In a saucepan, melt butter and blend in flour. Cook 1 minute; stir in beef broth, tomatoes and the reserved liquid, brandy, pepper, meat paste, and salt. Cook and stir. Serve over meat.

SWISS CREAM STEAK

Serves: 8

6 tablespoons margarine
2 cups onions, sliced
2 pounds tenderized round
  steak, cut into ½ inch strips
½ cup unsifted flour

1 tablespoon salt
1 teaspoon black pepper
1 teaspoon paprika
1 cup water
½ cup sour cream

Melt 4 tablespoons margarine in a large skillet. Add onions and sauté; remove from skillet. Dredge steak strips in flour seasoned with salt, pepper, and paprika. Melt remaining margarine; add steak to skillet. Brown meat well on both sides. Mix in onions, water, and sour cream until blended. Cover, cook over medium heat for 30 minutes. Uncover; cook until sauce thickens, approximately 10 minutes. Serve over cooked noodles or rice.

BREAKFAST BARS
Serves: 8

1 pound bulk sausage	4 eggs
1 medium onion, finely chopped	1 cup Bisquick
1 cup shredded Cheddar cheese	1 teaspoon salt
2 cups milk	Dash of pepper

Preheat oven to 400°.   13 x 9 baking dish

Lightly grease baking dish. Cook sausage and onions in a skillet; drain well. Place sausage and onions in baking dish. Sprinkle cheese over sausage mixture. Mix eggs, milk, Bisquick, salt, and pepper until smooth. Pour over sausage and cheese. Bake for 30-35 minutes until set and brown. Let set 5 minutes, then cut into bars.

BRUNCH CASSEROLE
Serves: 4-6

8 slices stale bread	8 eggs, beaten
½ cup butter	2½ cups milk
2 cups grated sharp cheese	½ cup half-and-half
1 (4 oz.) can green chilies, chopped	½ teaspoon salt
½ cup sliced green onions	¼ teaspoon cayenne pepper
8 slices Canadian bacon, quartered	¼ teaspoon dry mustard

Preheat oven to 325°.   13 x 9 baking pan

Lightly grease baking pan. Trim crusts off bread. Butter one side; cut into cubes. Layer bread, cheese, chilies, green onion, and Canadian bacon. Mix beaten eggs, milk, half-and-half, salt, cayenne pepper, and dry mustard. Pour over layered ingredients in pan. Bake uncovered for 45 minutes.

This can be made ahead and refrigerated before baking.

 Fresh egg shells are rough and chalky; old eggs are smooth and shiny.

BRUNCH EGGS OVER TOAST

Serves: 12-14

3 large green bell peppers, seeded and coarsely chopped
6 large onions, coarsely chopped
1-2 cloves garlic, finely minced
1 pound bacon, fried and diced
1 tablespoon flour
1 pound Longhorn cheese, grated

1 (14½ oz.) can tomatoes, undrained
½ small can diced jalapeños, undrained
1 dozen hard cooked eggs, cut into eighths
Toast rounds or Ritz crackers

Sauté peppers, onions, and garlic in bacon drippings. Remove; add flour to drippings. Stir until smooth. Add juices, tomatoes, and cheese. Add eggs, bacon, and jalapeños; stir gently. Serve over toast or crackers.

WINE AND CHEESE OMELET

Serves: 12-16

1 large loaf day-old French or Italian bread, crumbled
6 tablespoons unsalted butter, melted
¾ pound Swiss cheese, shredded
½ pound Monterey Jack cheese, shredded
9 thin slices Genoa salami, chopped
16 eggs

3¼ cups milk
½ cup dry white wine
4 large green onions, minced
1 tablespoon whole grain mustard
¼ teaspoon freshly ground black pepper
⅛ teaspoon ground red pepper
1½ cups sour cream
⅔-1 cup freshly grated Parmesan or shredded Asiago cheese

Preheat oven to 325°.   2 (13 x 9) baking pans

Lightly butter baking pans. Spread bread over bottom of baking dishes; drizzle with butter. Sprinkle with Swiss and Monterey Jack cheeses and salami. Beat eggs, milk, wine, onion, mustard, and peppers together until foamy. Divide and pour into baking dishes; cover with foil. Refrigerate overnight, or up to 24 hours. Remove from refrigerator 30 minutes before baking. Preheat oven to 325°. Bake covered casserole about 1 hour or until set. Uncover; spread with sour cream and sprinkle with Parmesan or Asiago cheese. Bake uncovered until crusty and light brown, approximately 10 minutes.

Can substitute bacon or ham for salami.

HAM AND CHEESE FRENCH TOAST

Serves: 12

12 1-inch thick slices firm,
   coarse textured white bread
   with crusts removed
3 cups whole milk
3 eggs, beaten
3 tablespoons sugar
3 tablespoons brandy or
   bourbon
1 teaspoon grated lemon peel
1 teaspoon grated orange peel
1 teaspoon freshly ground
   pepper

½ teaspoon salt
¼ teaspoon freshly grated
   nutmeg
12 (1 oz.) slices sharp Cheddar
   cheese, preferably white
6 (2 oz.) slices baked ham
4 tablespoons unsalted butter
4 tablespoons (or more)
   vegetable oil
Powdered sugar
Watercress sprigs
Warm Apricot Syrup

Dry bread slices at room temperature on wire rack for 1 hour.

Whisk next 9 ingredients in large bowl until smooth. Pour batter into 2 large rectangular baking dishes. Arrange 6 bread slices in each dish. Let stand until batter is absorbed, at least 1 hour; do not turn bread over. (Can be prepared 1 day ahead, covered, and refrigerated.) Cut cheese and ham slices to fit bread. Set 6 cheese slices on 6 bread slices. Top with ham. Cover with remaining cheese. Using spatula, place remaining bread dry side down atop ham. Melt 2 tablespoons butter with 2 tablespoons oil in each of 2 heavy skillets over medium heat. Place 3 sandwiches in each skillet and fry until outsides are crisp and golden brown and cheese has melted, adding more oil if necessary, about 5 minutes per side. Transfer to platter. Dust with powdered sugar. Garnish with watercress. Serve immediately with syrup.

Warm Apricot Syrup:

1 (12 oz.) jar apricot preserves,
   larger pieces of fruit cut up
3 tablespoons brandy or
   bourbon

2 tablespoons superfine sugar
1½ tablespoons fresh lemon juice
1 tablespoon water
¼ teaspoon grated lemon peel

Cook all ingredients in heavy small saucepan over low heat until preserves are melted and flavors blended, stirring occasionally, about 5 minutes.

 To maintain high quality in eggs, always store eggs large end up in their original container.

DIETER'S QUICHE

Serves: 8-10

10 eggs
1 (16 oz.) carton cottage
   cheese
1 pound Monterey Jack cheese,
   grated
2 (4 oz.) cans green chilies,
   chopped

1 (4.5 oz.) jar chopped
   mushrooms
½ cup unsifted flour
1 teaspoon baking powder
¼ cup butter or margarine

Preheat oven to 350°.   13 x 9 baking pan

Beat eggs until lemon colored. Mix all ingredients together and beat until well blended. Pour into glass dish or two small glass dishes. Bake for 35 to 45 minutes. Cool 5 minutes before slicing. May be served hot or cold.

Great with hot sauce or fresh fruit on the side.

RANCH-STYLE SAUSAGE AND GRITS

Serves: 8

1 cup quick-cooking grits
1 pound bulk pork sausage
1 large chopped onion
1 (7½ oz.) can salsa ranchero

½ pound Cheddar cheese,
   shredded and divided into 2
   parts

Preheat oven to 350°.   10 x 6 baking pan

Lightly grease baking dish. Cook grits according to instructions on box; set aside. Crumble sausage in a large skillet; add onion. Cook over medium heat until meat is browned, stirring occasionally. Drain well. Combine grits, sausage mixture, salsa, and half of the cheese. Spoon into baking pan. Bake for 15 minutes. Sprinkle with remaining cheese, and bake an additional 5-10 minutes, or until cheese melts.

 Eggs will beat up fluffier if they are allowed to come to cool room temperature before beating.

SPINACH AND CHEESE PIE

Serves: 12-15

1 large onion, finely chopped
3 tablespoons vegetable or olive oil
4 (10 oz.) packages chopped frozen spinach, defrosted and well drained
1 cup grated mozzarella (or white Cheddar) cheese

½ cup butter, melted
3 eggs
4 tablespoons milk
Salt and pepper, to taste
1 pound phyllo pastry sheet, thawed according to instructions on package

Preheat oven to 350°.   12 x 8 baking pan

Lightly butter baking pan. Sauté onion in oil for about 5 minutes; remove from heat. Mix together the spinach, cheeses, eggs, milk, salt, pepper, and onions. In baking pan, lay 1 sheet of pastry and brush sparingly with a little butter; take a second sheet, lay on top of first, and butter in same manner. Do this until half of the sheets have been used (10-12).

Spread all of spinach mixture over layered phyllo. Next, layer on remaining phyllo sheets with butter as with the first half. Butter the top of the last sheet, then bake for 30 minutes, or until pie is golden brown. Cut into squares before serving.

Best if topped with a yogurt sauce or plain yogurt.

SAVORY SPINACH QUICHE

Serves: 8-10

1 (10 oz.) package frozen chopped spinach
1 (3 oz.) package cream cheese, softened
1 cup shredded sharp Cheddar cheese
5 eggs, slightly beaten
¼ teaspoon salt

¼ cup chopped green onion
1 minced garlic clove (optional)
2 tablespoons chopped parsley
1 (9 inch) pastry shell, unbaked
1 thinly sliced tomato
¼ cup grated Parmesan cheese

Preheat oven to 450°.   9 inch pie pan

Cook spinach according to package directions. Drain well and squeeze to remove excess water. Set aside. Combine cream cheese, Cheddar cheese, eggs, salt, onion, garlic (optional), and parsley; beat lightly with a fork. Stir in spinach and pour above mixture into unbaked pastry shell. Arrange tomato slices on top and sprinkle with Parmesan cheese. Bake for 35 minutes or until set.

SUNRISE EGG CASSEROLE

Serves: 6-8

4 eggs
2 cups milk
¾ pound sausage, browned

¾ cup grated Velveeta cheese
6 slices bread, trimmed and
cubed

Preheat oven to 350°.   13 x 9 baking pan

Lightly butter baking pan. Beat eggs; add milk, sausage, and cheese. Pour over bread and mix well. Pour into prepared pan and cover with foil. Bake 20 minutes. Remove foil and turn oven up to 375° for 5-10 minutes more.

This can be prepared the night before; refrigerated, then baked in the morning.

GRILLED DOVE

Cleaned and breasted doves
1 fresh onion, cut into 1 inch
pieces

Pickled jalapeño slices
Thinly sliced bacon strips

Between the joined dove breasts, place one piece of onion and a jalapeño slice (to taste). Wrap each dove breast with a bacon strip. Use toothpicks to hold dove breast and bacon together. Grill over coals (low heat) for 10-20 minutes.

Add a 1 x 1 inch chunk of Monterey Jack cheese for a variation.

MARINATED DOVES

Serves: 10-12

30-40 doves   1 package shrimp boil

Preheat oven to 350°.   13 x 9 baking pan

Boil doves in shrimp boil for approximately 45 minutes to remove wild flavor; drain. Pour "Barbecue Sauce for Birds" over doves and marinate for 3 to 4 hours (or overnight) in refrigerator. Bake birds in sauce 30 minutes, or until lightly browned.

Barbecue Sauce for Birds:
1 pint white wine vinegar
1 (5 oz.) bottle Worcestershire
sauce
1 cup vegetable oil

½ cup sugar
Dash of salt and pepper
Dash of Tabasco sauce

Mix above ingredients together in a saucepan. Boil for 3 to 5 minutes.

STUFFED WHITEWING DOVE BREASTS

Serves: 6-8

30 whitewing doves
6 ounces Cheddar cheese

5 pickled jalapeño peppers
15 slices of bacon

Preheat oven to 350°.   13 x 9 baking pan

Remove skin and legs from doves; carefully filet each dove breast. Cut cheese into 30 strips. Wash and seed jalapeño peppers; slice each pepper into 6 strips. Stuff each dove breast with a slice of pepper and a slice of cheese. Wrap in ½ slice of bacon; secure with a toothpick. Bake 15 to 20 minutes, turn doves, and bake 10 minutes more, or until bacon is crisp.

WHITEWING BREASTS IN WINE SAUCE

Serves: 6

18 skinned whitewing breasts
½ cup flour
   Salt and pepper
2 tablespoons oil

2 (10½ oz.) cans cream of
   mushroom soup
1⅓ cups dry white wine
   Cooked white rice

Preheat oven to 300°.   13 x 9 baking dish

Dredge breast in seasoned flour. Brown on both sides in hot oil in a large skillet. Arrange breasts in a single layer in the dish. Mix soup and wine together; pour over breasts. Cover and bake for 1½ hours. Remove cover and bake 15 minutes longer. Serve over rice.

GRILLED DUCK

Serves: 1-2 per duck

   Clean picked ducks
   Olive oil
   Black pepper
   Cayenne pepper

   Onion, cut into chunks
   Apple, sliced
   Celery, chopped
   Bacon

Start fire in barbecue pit, allow coals to burn low.

Brush ducks with olive oil. Sprinkle peppers, inside and out. Stuff cavity with apple, celery, and onion. Secure bacon around ducks with toothpick. Place on grill and cover. Ducks are done when the bacon is cooked.

ROASTED DUCK

Serves: 3-4

6 slices bread, cubed and
toasted
2 stalks celery with leaves,
chopped
4 tablespoons chopped onions
4 tablespoons slivered almonds
Garlic powder to taste
2 tablespoons chervil

2 tablespoons thyme
4 tablespoons melted margarine
2 ducks
1½ cups chicken broth
1½ cups white wine
Margarine
Flour

Preheat oven to 325°.   13 x 9 baking pan

Combine bread, celery, onions, almonds, garlic powder, chervil, thyme; stuff
into ducks. Brush ducks with melted margarine. Place in roaster; add broth and
wine; bake covered for 4 hours. Add the pan juices to a roux made of margarine
and flour. Stir until thick. Serve this gravy with the ducks and stuffing.

GOOSE BREAST

Serves: 6-8

6-8 boneless breasts of goose

Marinade:

1 cup Apple Jack (brandy can
be substituted)
½ cup olive oil
10 cloves garlic, finely chopped
1 medium onion, chopped

3-4 bay leaves
⅛ teaspoon sage
⅛ teaspoon pepper, coarsely
ground
⅛ teaspoon thyme

Combine all of the above ingredients, except for the goose breasts. Marinate the
goose breasts for at least 4 hours. Grill the breasts over hot coals for 6 to 10
minutes (3 to 5 minutes per side). The breasts should be RARE to MEDIUM-
RARE, so they will be juicy. (Wild goose tends to be tough and tasteless when
overdone.)

This marinade is also good for venison!

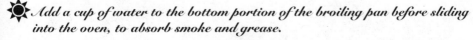 *Add a cup of water to the bottom portion of the broiling pan before sliding
into the oven, to absorb smoke and grease.*

QUAIL IN ORANGE SAUCE

Serves: 4-5

8-10 quail, cleaned and skinned
Melted butter (or margarine)
1 cup orange marmalade
¼ cup brown sugar
3 tablespoons wine vinegar
2 teaspoons Worcestershire
sauce

½ teaspoon curry powder
½ teaspoon ground ginger
1 teaspoon monosodium
glutamate
Generous dash of cayenne
pepper
Salt and pepper to taste

Preheat oven to 350°.   13 x 9 baking pan

Wash and dry quail. Brush with melted butter. Combine remaining ingredients in saucepan; cook over medium heat until boiling. Simmer for 2 minutes, stirring constantly. Remove from heat; brush quail with sauce. Roast in oven for 20 minutes. Turn quail; roast for 20 minutes more. Brush with sauce throughout roasting. Serve immediately.

VENISON STEAK FROMAGE

Serves: 4-6

Venison steak or backstrap
⅓ cup flour
½ teaspoon salt
⅛ teaspoon pepper
¼ teaspoon garlic salt

3 tablespoons oil
⅓ cup onion, chopped
¾ cup water
⅓ cup Cheddar cheese, grated
2 tablespoons parsley

Cut venison into serving size pieces; tenderize. Combine flour, salt, pepper, and garlic salt. Dredge steak in flour mixture; brown in hot oil. Add onions and water; sprinkle remaining flour mixture over the top. Cover; simmer for ½ hour. Remove cover. Sprinkle on grated cheese and parsley.

 When pan frying, always heat the pan before adding the butter or oil.

HOT AND SPICY VENISON

Serves: 6-8

1 pound venison, back strap

Cut venison, across the grain, into very thin pieces.

Meat Marinade:

3 tablespoons soy sauce
(preferably China brand)
4 teaspoons sherry

2 tablespoons cornstarch
4 teaspoons sesame oil
Dash of white pepper

Soak the venison in the marinade for at least ½ hour.

Sauce:

4 tablespoons fresh minced
ginger
4 tablespoons fresh minced
garlic
4 tablespoons hot bean sauce*
4 tablespoons sweet bean*
4 tablespoons soy sauce
4 tablespoons sherry
2 teaspoons rice vinegar*

4 teaspoons sugar
2 teaspoons cornstarch
1 large green bell pepper, thinly
sliced
1 large onion, cut into small
cubes
4-5 whole dried red chili peppers
3 tablespoons peanut oil,
divided

Combine sauce ingredients in a separate bowl, mixing well. Heat wok, adding 1 tablespoon of oil. Stir fry green pepper with onion for 2 minutes. Add a pinch of salt and ¼ teaspoon of pepper to taste, set aside. Add 1-2 tablespoons of oil, and add whole dried red chili pepper. Let cook for 20 seconds, and add venison. Stir fry venison for 1 minute. Add sauce, and stir until thick. Add the vegetables back into the wok just long enough to reheat. Serve over rice.

(Venison should be cut like a flank steak. When thinly sliced across the grain and cooked quickly, it becomes juicy and tender.)

*You can purchase these products in an Oriental market.

 To remove a stubborn cork, wrap a cloth dipped in boiling water around the neck of the bottle. After a few minutes, the cork will come out easily.

LAMB ALEXANDER

Serves: 4

Garnish:

8 dried apricots   ½ cup Courvoisier

Simmer the dried apricots in a small saucepan in Courvoisier until liquid is absorbed (approximately 10 minutes). Reserve.

Meat:

8 lamb chops, 1-inch thick	½ cup Madeira wine
1 tablespoon paprika	1 cup Courvoisier
2 tablespoons butter	½ pint whipping cream
2 tablespoons soy sauce	8 teaspoons apricot jam

Trim the excess fat from the lamb chops and coat them with paprika. Melt butter in a large skillet; brown the lamb chops on both sides. Pour off any excess fat. Sprinkle soy sauce on top of the chops. Pour in the wine; simmer for about 5 minutes until the liquid begins to reduce. Warm the Courvoisier, pour over the chops, and ignite. When the flames have died down, remove the chops; add the whipping cream to the sauce. Boil to reduce the sauce to 1 cup. Return the chops to the sauce; top each with 1 teaspoon of apricot jam. Heat. To serve, place the chops on a platter, pour the sauce over each, and then top each with one of the dried apricots simmered in Courvoisier.

LINGUINE WITH WHITE CLAM SAUCE

Serves: 4

3 (8 oz.) cans minced or whole baby clams	2 tablespoons white wine
	1 teaspoon fresh basil
¼ cup olive oil	½ teaspoon salt
3 garlic cloves, minced	1 pound fresh or refrigerated
¾ cup chopped parsley	linguine

Drain juice from clams, reserving juice. Heat olive oil in 2 quart saucepan over medium heat; add garlic and cook until tender. Stir in reserved clam juice and remaining ingredients except the clams. Cook 10 minutes, stirring occasionally. Stir in drained clams. Cook sauce until clams are heated through. Cook linguine in boiling salted water about 3 minutes or according to package instructions. Serve the linguine topped with the sauce.

Serve with warm bread and Pinot Blanc wine.

PASTA PRIMAVERA WITH MARINARA SAUCE

Serves: 4

Marinara Sauce:

4⅔ cups tomato sauce
⅔ quart water
2 cups finely chopped onion
1½ teaspoons finely chopped garlic

½ teaspoon oregano
½ teaspoon rosemary
½ teaspoon thyme
1 bay leaf
¼ teaspoon black pepper

Combine all ingredients; bring to a boil. Reduce heat; simmer uncovered for at least 2 hours.

Pasta Primavera:

1½ tablespoons olive oil
1 garlic clove, minced
2 pounds fettuccine, cooked al dente and drained
4 cups fresh vegetables

2 cups marinara sauce
1 cup Parmesan cheese, freshly grated
½ cup pine nuts, toasted
Fresh basil, chopped

Combine olive oil and garlic. Toss cooked pasta in oil. Steam vegetables until tender but crisp; rinse in cold water to maintain color. Blanch to reheat just before serving. For each serving, combine 1 cup pasta, ½ cup marinara sauce, 1 cup vegetables, ¼ cup cheese, 2 tablespoons pine nuts, and basil. Serve on hot plate; garnish with a fresh basil leaf.

LIGHT FETTUCCINE

Serves: 6-8

1 cup light margarine
1 (8 oz.) container light sour cream
¼ teaspoon paprika
¼ teaspoon pepper
¼ teaspoon poultry seasoning
Garlic powder to taste

1 teaspoon dried parsley
2 tablespoons grated Parmesan cheese
½ cup lowfat milk, heated
12 ounces thin egg noodles, cooked and drained

Combine all ingredients, except milk and noodles, and heat until cheese and butter are melted. Add milk and noodles. Serve immediately.

ITALIAN STUFFED SHELLS

Serves: 10

12 ounces jumbo shells
1 pound hamburger meat
1 (48-56 oz.) jar spaghetti
 sauce
4 cups ricotta cheese
2 cups shredded mozzarella
 cheese

¾ cup grated Parmesan cheese
3 eggs
½ cup chopped fresh spinach
½ teaspoon salt
½ teaspoon pepper

Preheat oven to 350°.   2 (13 x 9) baking pans

Cook shells according to package directions and cool in a single layer on aluminum foil to prevent sticking. Brown hamburger in a large skillet; drain grease and add spaghetti sauce to meat. Line pans with a thin layer of meat sauce. Mix together cheeses, eggs, spinach, salt, and pepper. Fill each shell with a heaping tablespoon of cheese mixture. Squeeze shell slightly and place open side down in pan. Cover with remaining sauce. Bake uncovered for 35 minutes.

PERFECT PORK TENDERLOIN

Serves: 6-12

2 pork tenderloins

Marinade:
⅔ cup soy sauce
⅔ cup oil
2 tablespoons crystallized
 ginger, finely chopped
2 tablespoons lime juice

1 teaspoon garlic powder
2 tablespoons minced onion
1 tablespoon monosodium
 glutamate

Combine all ingredients of the marinade; pour over the pork tenderloin. Marinate 36 hours in a resealable plastic bag or in a container of your choice, turning meat occasionally. When ready to serve, grill over charcoal about 45 minutes or on a gas grill for about 30 minutes over medium heat, basting occasionally with remaining marinade. Do not overcook.

 To remove a stubborn set-in crease in table linens, spray vinegar on crease, then iron it.

SESAME PORK TENDERLOIN

Serves: 6

Marinade:
- ¼ cup reduced sodium soy sauce
- ¼ cup orange juice
- 1 tablespoon peeled, minced ginger
- 2 tablespoons honey
- 1 clove garlic, minced

Combine marinade in a large resealable plastic bag. Seal bag; shake well.

Meat:
- 2 (¾ lb.) pork tenderloins
- 3 tablespoons sesame seeds

Trim fat from the pork. Add pork to bag; seal bag and shake until pork is coated. Marinate in refrigerator 8 hours, or overnight, shaking bag occasionally. Remove pork from marinade; drain, reserving marinade. Sprinkle pork evenly with sesame seeds. Coat rack of a roasting pan with non-stick spray. Place pork on rack. Bake at 350° for 45-55 minutes. Remove pork from oven; let stand for 15 minutes.

Sauce:
- ¼ cup red wine
- ¼ cup chicken broth
- 2 tablespoons water
- 2 teaspoons cornstarch

Strain reserved marinade into a medium saucepan. Add red wine and chicken broth, stirring well. Bring mixture to a boil. Reduce heat; simmer 5 minutes. Combine water and cornstarch; stir until smooth. Add this to chicken broth mixture, stirring well. Cook until sauce is slightly thickened, stirring constantly. Slice pork diagonally into thin slices. Spoon sauce over pork.

 Add beer to meats you boil, braise, or roast. It will act both as a tenderizer and a flavor enhancer.

COMPANY PORK
Serves: 4

1½-2 pounds pork tenderloin, cut
   in 1-inch thick medallions
2 cups pineapple juice

½ cup soy sauce
2 garlic cloves, crushed

Place pork medallions in dish. Combine remaining ingredients; pour over pork. Cover and refrigerate overnight, turning once. Cook on medium grill, basting frequently with marinade.

You can cook meat ahead; put back in marinade, cover, and hold in oven on warm until ready to serve.

CHICKEN WITH BING CHERRY SAUCE
Serves: 4

1 frying chicken, cut up
2 cups lemon juice

Salt
Pepper

Preheat oven to 350°.

Place chicken pieces in bowl to marinate in lemon juice, salt, and pepper. Marinate 1-2 hours. Place chicken on broiler pan. Broil until pieces are brown, and then bake for 1 hour.

Bing Cherry Sauce:

1 medium onion, grated
4 tablespoons butter
2 tablespoons cornstarch
4 tablespoons sugar
   Juice from one can of Bing
   cherries

½ cup port wine
¼ cup cherry brandy
Dash of salt
Dash of pepper
1 (15 oz.) can of Bing cherries

Sauté onion in butter until limp. Combine cornstarch with sugar and cherry juice. Stirring constantly, bring to boil; allow mixture to thicken about 3 minutes. Stir in port wine, cherry brandy, salt, and pepper. Add Bing cherries, simmer 2-3 minutes. Pour over chicken and bake 5 minutes. Serve over wild or regular rice.

 Tomato juice used to soak a hopelessly black frying pan for 30-45 minutes will let you wash it clean as new.

CHICKEN WITH BROCCOLI CORNBREAD

Serves: 9

Cornbread:

1   (10 oz.) package frozen
    chopped broccoli, thawed
½ cup margarine
1   teaspoon salt

6   ounces cottage cheese
1   large onion, chopped
4   eggs, well beaten
1   box Jiffy cornbread mix

Preheat oven to 400°.   13 x 9 baking pan

Drain water off broccoli. Mix all ingredients together; add cornbread mix last. Bake 30 minutes.

Creamed Chicken:

2   (6 oz.) cans sliced
    mushrooms
1   cup diced green bell pepper
1   cup margarine
1   cup flour
2   teaspoons salt
½ teaspoon pepper

2   cups half-and-half, (or whole
    milk)
2½ cups chicken broth
4   cups chopped cooked chicken
2   (4 oz.) jars pimientos,
    chopped and drained

Drain mushrooms, reserving ½ cup of liquid. In a Dutch oven, cook mushrooms and pepper in margarine for 5 minutes. Stir in flour, salt, pepper, and liquids. Cook over low heat, stirring constantly until sauce boils. Boil 1 minute. Stir in chicken and pimiento; heat until hot. Serve over Broccoli Cornbread.

CHICKEN CURRY CASSEROLE

Serves: 6-8

2   cups Minute Rice, uncooked
1   large head broccoli, tops cut
    into bite-size pieces
1   pint mushrooms, sliced
6   chicken breasts, cut into
    bite-size pieces
2   cans (10¾ oz.) cream of
    chicken soup

2   (13¾ oz.) can chicken broth
¾ cup mayonnaise
2   teaspoons curry powder
½ teaspoon pepper
2   tablespoons lemon juice

Preheat oven to 350°.   13 x 9 baking pan

Prepare baking pan. Layer rice, broccoli, mushrooms, and chicken in pan. In large mixing bowl, combine soups, mayonnaise, curry powder, pepper and lemon juice. Mix well. Pour soup mixture over top of layers. Bake uncovered for 45 minutes.

CHICKEN KIEV

Serves: 4

4 chicken breasts, skinned and
  boned
½ cup butter
1 tablespoon chopped parsley
1 tablespoon chopped chives
  Salt and pepper

1 cup flour
1 egg, beaten with 1 tablespoon
  water
Dried bread crumbs
Deep fat or oil for frying

Pound chicken between 2 sheets of waxed paper until thin. Cream butter until soft; beat in parsley and chives. Divide butter mixture into 4 servings; place in the middle of each chicken breast; salt and pepper it. Roll chicken breast in envelope fashion; secure with 3-4 toothpicks. Toss envelopes in flour seasoned with salt and pepper; dip in egg mixture; roll in bread crumbs. Chill in refrigerator 1 hour. Fry in deep hot fat or oil for 12 minutes.

Make sure envelope is tight with toothpicks so the butter does not escape while frying.

CHICKEN MARSALA

Serves: 4

4 chicken breasts, skinned and
  boned
1 cup flour
¼ teaspoon salt
¼ teaspoon white pepper

½ cup butter
2 cups sliced fresh mushrooms
1 cup Marsala wine
Parsley for garnish

Pound chicken breasts to ¼ inch thickness. Coat chicken with mixture of flour, salt, and white pepper. Melt ¼ cup butter in skillet; sauté mushrooms. Remove mushrooms, add remaining butter, and brown chicken on both sides. Return mushrooms to skillet; add wine. Simmer until chicken is done and sauce thickens. Garnish with fresh parsley.

After flouring chicken, chill for one hour. The coating adheres better during frying.

CHICKEN PARMESAN

Serves: 6-8

½ cup vegetable oil
2 eggs, lightly beaten
1 teaspoon salt
Dash of pepper
4 whole chicken breasts, split, skinned, and boned
1 cup fine, dry, unseasoned bread crumbs
4 cups tomato sauce

½ teaspoon basil
½ teaspoon oregano
1 tablespoon parsley
Dash of garlic powder
2 tablespoons butter
½ cup grated Parmesan cheese
8 ounces mozzarella cheese, sliced

Preheat oil in skillet to 360°.   11 x 8 baking pan

Preheat oven to 350°. Combine eggs, salt, and pepper. Dip chicken breasts in egg mixture, then into bread crumbs. Heat oil in electric skillet. Brown chicken in skillet on both sides; remove to casserole dish. Mix together tomato sauce and herbs in a saucepan. Heat to boiling; reduce heat, and simmer 10 minutes. Add butter; when melted, pour sauce over chicken. Sprinkle with Parmesan cheese, cover, and bake for 30 minutes. Uncover, place mozzarella cheese on top, and bake 10 minutes longer.

CHICKEN BREASTS WITH PESTO SAUCE

Serves: 4

4 chicken breasts (bone in or boneless)
3 tablespoons fresh lemon juice
¼ teaspoon salt
⅛ teaspoon freshly ground pepper

¼ cup minced fresh basil
¼ cup grated Parmesan cheese
1 garlic clove, crushed through a press
2 tablespoons olive oil

Preheat broiler.   Broiler pan

Brush chicken breasts with lemon juice; season with salt and pepper. Place chicken on broiler pan; broil approximately 8 inches from heat for 5 minutes. Turn and broil 5 minutes longer. Meanwhile, combine basil, cheese, garlic, and oil in blender. Mix until well blended. Spread over each chicken breast; broil 5 to 10 minutes longer, or until chicken is opaque throughout, but still juicy.

CHICKEN BREASTS PROSCIUTTO

Serves: 6

3 large whole chicken breasts,
   skinned and boned
¼ cup flour
   Dash of garlic powder
   Dash of white pepper
4 tablespoons butter

2 ounces Prosciutto slices
4 ounces Jarlsburg cheese,
   thinly sliced
½ cup white wine
½ cup chicken broth
1 tablespoon brandy

Preheat oven to 350°.   2 quart shallow baking dish

Slice 2-3 fillets from each side of breasts. Combine the flour, garlic, and pepper; lightly dredge the fillets. Heat butter in a heavy skillet. Cook fillets slowly over low to medium heat for 2-3 minutes. Do not overcook. Place chicken in baking dish. On each piece of chicken, arrange a thin slice of Prosciutto; top with a thin slice of Jarlsburg cheese. Add wine, chicken broth, and brandy to juices in pan. Simmer until the liquid is reduced and slightly thickened, about 5 minutes. Pour over breasts. (At this point it may be refrigerated up to 48 hours.) Bake, uncovered, for 15-20 minutes or until hot and bubbly.

CHICKEN SPECTACULAR

Serves: 6-8

1 whole chicken or 4 large
   breasts
1 (6 oz.) package long grain
   and wild rice
1 (2 oz.) package slivered
   almonds
   Butter and salt
1 (1 lb.) can French style green
   beans, drained

1 (10¾ oz.) can cream of celery
   soup
½ cup mayonnaise
1 cup drained and sliced water
   chestnuts
2 tablespoons chopped
   pimientos
1 (6 oz.) can sliced mushrooms
2 tablespoons chopped onion

Preheat oven to 350°.   2 quart casserole

Stew chicken; dice. Use the broth from stewing to cook rice. Toast almonds in butter and salt under broiler. Mix beans, soup, mayonnaise, chestnuts, pimiento, mushrooms, onion, and almonds with chicken and rice. Place in casserole and bake, uncovered, for 30 minutes.

CHICKEN WITH YORKSHIRE PUDDING

Serves: 6

Baked Chicken:

1 (3 lb.) frying chicken, cut up
⅓ cup flour
½ teaspoon monosodium glutamate

1 teaspoon salt
1 teaspoon paprika
½ cup melted butter

Preheat oven to 400°.   13 x 9 baking pan

In brown bag combine flour, salt, monosodium glutamate, and paprika. Shake bag to coat pieces well with mixture. Put into baking pan; pour butter evenly over chicken. Bake for 30 minutes. Remove from oven; pour off drippings, reserving ¼ cup.

Yorkshire Pudding:

3 eggs
1 cup milk
½ cup sour cream
½ teaspoon salt
¼ cup drippings (reserved from baked chicken)

1 cup flour
¼ teaspoon baking powder
¼ teaspoon poultry seasoning
Chopped parsley (optional)

Beat eggs, milk, and sour cream together with rotary beater. Add rest of ingredients. Beat until smooth. Pour mixture over chicken; bake 30 minutes until Yorkshire pudding is puffed and browned. Sprinkle top with chopped parsley, if desired.

CRANBERRY CHICKEN

Serves: 6

6 chicken breast halves, skinned
1 (16 oz.) can whole cranberry sauce

1 (8 oz.) bottle French dressing
1 package Lipton's onion soup mix

Preheat oven to 350°.   13 x 9 baking pan

Grease pan. Place chicken breasts in pan. Mix remaining ingredients together; pour over chicken. Bake covered for 1 hour.

165

CITRUS TARRAGON CHICKEN

Serves: 4

8 chicken breast halves,
skinned and boned

2 teaspoons cornstarch
dissolved in 1 tablespoon
water

Marinade:

1 tablespoon grated orange peel
¼ cup fresh lemon juice
2 tablespoons Worcestershire
sauce
1 tablespoon fresh tarragon (or
1 teaspoon dried tarragon)

1 cup fresh orange juice
⅓ cup honey
1 teaspoon dry mustard
Salt and pepper (to taste)

Preheat oven to 350°.   13 x 9 baking pan

Combine all marinade ingredients, pour over chicken, cover, and refrigerate 2 hours. Place chicken and marinade in large flat baking dish. Bake covered for 30 minutes. Take chicken out of marinade. In saucepan combine marinade with cornstarch; cook over low heat, stirring until thickened. To serve, ladle sauce over chicken.

CURAÇAO CHICKEN

Serves: 6

1 whole chicken, cut up
  Comino (Cumin)
  Black pepper
  Olive oil
3 tablespoons butter
3 cloves garlic, minced
1 bell pepper, sliced
1 medium onion, sliced

1 (14½ oz.) can whole
  tomatoes
1 bay leaf
3 tablespoons Worcestershire
  sauce
3 tablespoons catsup
2 teaspoons cider vinegar

Season chicken lightly with comino and pepper. Brown chicken in olive oil (enough to coat ¼ inch of frying pan) over medium heat. Remove chicken and all but 3 tablespoons of oil. Add butter; over medium heat sauté garlic, pepper, and onion until tender. Add tomatoes, bay leaf, Worcestershire sauce, catsup, vinegar; cook 1 minute. Add chicken pieces; spoon sauce on top. Cook 20 minutes over low heat.

HONEY CRUNCH CHICKEN

Serves: 4

4 chicken breasts, halved and
   skinned
2 tablespoons reduced calorie
   mayonnaise

2 tablespoons nonfat yogurt
1½ ounces Grape Nuts cereal
1¼ tablespoons honey

Preheat oven to 350°.   13 x 9 baking pan

Wash chicken; pat dry. Place in baking pan. Mix mayonnaise and yogurt.
Spread mixture over both sides of chicken with a pastry brush. Crush cereal in
a blender or food processor. Sprinkle crushed cereal evenly over top side of
chicken. Drizzle chicken evenly with honey. Bake uncovered 40-45 minutes.

IMPERIAL CHICKEN

Serves: 8

¾ cup butter, melted
2 cloves garlic, pressed
1 cup fine dry bread crumbs
⅔ cup grated Parmesan cheese
¼ cup fresh parsley, minced
1 teaspoon salt

¼ teaspoon pepper
4 whole chicken breasts, split,
   skinned, and boned
Juice of 2 lemons
Paprika

Preheat oven to 350°.   13 x 9 baking pan

Grease pan. Combine butter and garlic; stir well; set aside. Combine bread
crumbs, cheese, parsley, salt, and pepper; stir well. Dip chicken in butter
mixture and coat with bread crumb mixture. Starting at small end, roll chicken;
secure with toothpick, if desired. Place chicken rolls in pan; sprinkle with lemon
juice and paprika. Bake for 45 minutes.

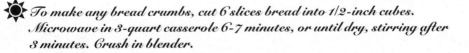

*To make any bread crumbs, cut 6 slices bread into 1/2-inch cubes.
Microwave in 3-quart casserole 6-7 minutes, or until dry, stirring after
3 minutes. Crush in blender.*

MONTEREY CHICKEN AND RICE

Serves: 6

½ cup cottage cheese
1 (3 oz.) package cream cheese
½ cup sour cream
1 (10¾ oz.) can cream of
   chicken soup
1 teaspoon salt
⅛ teaspoon garlic powder

1 (4 oz.) can diced green chilies
3 cups cooked, diced chicken
3 cups cooked rice
1 cup grated Monterey Jack
   cheese
¾ cup coarsely crushed corn
   chips

Preheat oven to 350°.   2 quart baking dish

Blend cottage cheese, cream cheese, and sour cream until smooth. Add mixture to remaining ingredients, except corn chips. Pour into baking dish. Sprinkle with corn chips. Bake for 25 to 30 minutes.

OVEN FRIED SESAME CHICKEN

Serves: 4

3 tablespoons sesame seeds
2 tablespoons all-purpose flour
¼ teaspoon pepper
4 chicken breast halves,
   skinned

2-3 tablespoons soy sauce
1-2 tablespoons reduced-calorie
    margarine, melted

Preheat oven to 400°.   13 x 9 pan

Combine sesame seeds, flour, and pepper. Dip chicken pieces into soy sauce; dredge in sesame seed mixture. Arrange chicken, bone side down, in pan; drizzle margarine over chicken. Bake for 40 to 45 minutes or until chicken is tender.

ORANGE CHICKEN WITH PESTO SAUCE

Serves: 8

Chicken:

8 chicken breast halves, skinned and boned
Salt and freshly ground pepper to taste
1 cup fresh orange juice

12 rosemary sprigs, divided
10 orange slices
Rosemary-Orange Pesto (see recipe below)

Preheat oven to 375°.   13 x 9 baking pan

Rinse chicken in cold water and pat dry. Season chicken with salt and pepper; place in a single layer in baking pan. Cut a piece of parchment paper to fit snugly inside the pan; set aside. Pour orange juice over the chicken, tuck in 3 rosemary sprigs and top with parchment paper so it is touching the chicken. Cook 10 to 15 minutes, or until chicken is firm and opaque. Remove from oven and cool chicken in the poaching liquid. Drain and garnish with remaining rosemary and orange slices. Place a dab of pesto on top of each chicken breast and pass the remainder at the table.

Rosemary-Orange Pesto:

2 garlic cloves, crushed
2 tablespoons fresh rosemary
½ cup fresh parsley
1 tablespoon grated orange zest
1 cup coarsely chopped green onions
¼ teaspoon cayenne pepper

½ cup olive oil
2 tablespoons freshly grated Parmesan cheese
1 tablespoon balsamic vinegar
¼ cup coarsely chopped walnuts, toasted

In a blender or food processor fitted with the metal blade, combine and puree all the ingredients. Use immediately or cover and store in refrigerator up to 4 days. Freezes.

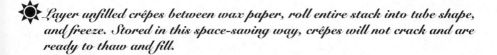

Layer unfilled crêpes between wax paper, roll entire stack into tube shape, and freeze. Stored in this space-saving way, crêpes will not crack and are ready to thaw and fill.

PEANUT CHICKEN ON SKEWERS

Serves: 6

6 chicken breast halves, skinned, boned, and cut into strips

Wooden skewers

Marinade:

½ cup chunky peanut butter
½ cup peanut oil
¼ cup white wine vinegar
⅓ cup soy sauce
⅓ cup lemon juice

4 cloves garlic, minced
8 cilantro sprigs
2 teaspoons red pepper flakes
2 teaspoons chopped fresh ginger

Combine marinade ingredients in food processor or blender. Blend well. Pour over chicken; refrigerate overnight. Soak skewers in water for 15 minutes. Skewer chicken. Grill over hot coals (or indoor grill) for 8-10 minutes, turning once.

PECAN CHICKEN

Serves: 2

½ cup flour
½ teaspoon salt
¼ teaspoon pepper
1 egg white, beaten until frothy
3 tablespoons grainy mustard, divided
½ cup finely ground pecans

½ cup fresh bread crumbs
2 (6 oz.) chicken breasts, skinned and boned
2 tablespoons butter
¼ cup heavy cream
Dash of pepper

In a shallow bowl, combine the flour, salt, and pepper. In a second bowl, blend the egg white with 2 tablespoons of the mustard. In a third shallow bowl, combine the pecans and bread crumbs. Dredge each chicken breast in the flour, shaking off excess. Next, dip into egg mixture to coat. Finally, dredge in pecan mixture, pressing so the coating adheres to both sides of the breast. Place the chicken on a rack; store uncovered in the refrigerator. In a medium skillet, melt the butter over medium-high heat. When the foam subsides, add the chicken breasts. Cook on one side until crisp and brown, about 3 minutes. Reduce heat to medium low, turn chicken, and continue cooking for about 5 minutes. Meanwhile, in a small saucepan, warm cream over medium heat. Add the remaining 1 tablespoon of mustard and a pinch of pepper; cook until thoroughly warmed. To serve, slice breasts on the diagonal and fan out. Serve sauce separately.

SAVORY YOGURT CHICKEN

Serves: 12

6 chicken breast halves,
skinned and boned
1 cup fine dry bread crumbs
¼ cup grated Parmesan cheese
1-2 tablespoons dried minced
onion
1 teaspoon garlic powder
1 teaspoon seasoned salt

¼ teaspoon dried oregano,
crushed
¼ teaspoon dried thyme
Dash of pepper
¼ cup butter or margarine,
melted
2 teaspoons sesame seeds

Preheat oven to 375°.

13 x 9 baking pan

Lightly grease pan. Mix bread crumbs, cheese, and spices (except sesame seeds) together in plastic bag. Coat chicken breasts with crumb mixture. Place chicken in pan. Drizzle butter on top; sprinkle with sesame seeds. Bake uncovered for 45 minutes.

Sauce:

1 (10¾ oz.) can cream of
chicken soup
1 (8 oz.) plain yogurt
½ cup chicken broth
1 teaspoon lemon juice

½ teaspoon Worcestershire
sauce
Dash of garlic powder
Dash of seasoned salt

In medium saucepan combine all ingredients; cook over low heat until hot. Serve over chicken.

VALLEY LEMON CHICKEN

Serves: 4-6

8 chicken breast halves,
skinned and boned
½ pound unsalted butter, melted

Juice of 1½ lemons
2 cups Italian bread crumbs

Preheat oven to 400°.

11 x 9 baking pan

Combine butter and lemon juice; reserve ¼ cup for sauce. Dip breasts into mixture on both sides. Roll in bread crumbs; bake uncovered for 35-40 minutes. Do not turn breasts. Place on serving platter; pour reserved lemon butter over chicken breasts.

TANDOORI CHICKEN

Serves: 8-10

8-10 chicken breasts, skinned

Garam Marsala:
- 2 tablespoons whole black peppercorns
- 1 tablespoon coriander seeds
- 1 teaspoon whole cardamom seed (pods removed)
- 4 teaspoons cumin seed
- 2 teaspoons whole cloves

Preheat oven to 300°.   Small baking pan

Place all spices in a baking pan; heat for 15 minutes. In a grinder, place heated spices; blend very fine (powder). Store the spice mixture in a tightly covered container in a cool, dry place. (Makes ⅓ cup.)

Marinade:
- 1 (8 oz.) carton plain yogurt
- 1 tablespoon red chili powder
- 1½ tablespoons ground cumin
- 2 tablespoons freshly ground black pepper
- 4 garlic cloves, minced
- 1 piece fresh ginger (walnut size)
- ½ cup white vinegar
- ½ cup oil
- 1 tablespoon salt
- ⅓ cup red food coloring
- 1 tablespoon Garam Marsala
- Lemon juice (only if grilled on coals)

Slash the chicken pieces 2 or 3 times with a sharp knife. Place in a large container. Mix all the marinade ingredients together; combine with the chicken. Marinate the chicken for a couple of hours (or refrigerate overnight).

Note: 1 heaping tablespoon garlic-ginger paste may be substituted for garlic and ginger.

Preheat oven to 500°.   13 x 9 baking pan

Place the marinated chicken (breast side up) in a shallow pan; roast until crusty, about 20 minutes. Turn over; roast another 10 minutes. The chicken should be deeply crusted, but never scorched.

(Continued on next page)

(Tandoori Chicken, continued from previous page)

To grill: Brush the chicken with the marinade; sprinkle with lemon juice. Grill until chicken is golden red, about 15 minutes. Turn and grill for an additional 10 minutes.

Be sure to use an abundance of charcoal, waiting until the bed has acquired a grey ash. To prevent sticking, wipe the grill with a little oil.

Garnish:

Lettuce Leaves   Coriander leaves
Onion and lemon slices
(paper thin)

Place chicken on a large platter on a bed of lettuce leaves. Garnish with paper-thin onion and lemon slices and coriander leaves.

ZESTY GRILLED CHICKEN

Serves: 6

2 large cloves garlic, finely chopped
1 large onion, grated
2 tablespoons dried parsley
¼ cup Dijon mustard
¼ cup lemon juice
¼ cup olive oil
Cracked pepper
1 tablespoon Worcestershire sauce
6 boneless chicken breast halves, skinned

Prepare outdoor grill or preheat broiler.   13 x 9 pan

Stir together well the garlic, onion, parsley, mustard, lemon juice, olive oil, pepper, and Worcestershire sauce; coat each chicken breast. Refrigerate 2 to 3 hours. Reserve marinade. Cook chicken 5 to 7 minutes on each side until browned and cooked through. Heat marinade on stove; simmer for 5 to 7 minutes. Pour a small amount of mixture over each chicken breast when serving.

CORNISH GAME HENS

Serves: 6

3 game hens
¼ cup olive oil
Juice of 2 or 3 limes
2 tablespoons minced fresh sage leaves
2 tablespoons chopped mint leaves

2 cloves garlic, minced
1 teaspoon salt
½ teaspoon freshly ground pepper
3 drops Tabasco sauce

Preheat oven to 450°.   13 x 9 glass baking dish

Split hens in half. Wash and dry well. Place in dish. Combine remaining ingredients; pour over hens. Cover; refrigerate 2-3 hours, turning hens in marinade several times. Uncover hens, roast in marinade 20 minutes. Reduce temperature to 400°. Bake 30-40 minutes more, or until hens are golden brown and tender, basting every 10-15 minutes. Transfer to serving platter, cover, and keep warm. Prepare Grape Sauce.

Grape Sauce:

1 cup chicken broth, heated
½ cup port or red wine
2 tablespoons cornstarch, dissolved in 2 tablespoons water

1 tablespoon brown sugar (or more to taste)
1 teaspoon fresh thyme or ¼ teaspoon dried thyme
2 cups seedless red grapes

Add hot chicken broth to juices in baking pan; stir to loosen brown particles. Pour into a saucepan; add wine, cornstarch mixture, sugar, thyme, and grapes. Cook over high heat, stirring until mixture thickens and clears. Pour half of sauce over hens and the rest into a sauce boat. Garnish platter with clusters of fresh grapes. Serve with rice.

For a spring menu, substitute green seedless grapes and white wine.

TURKEY DELIGHT

Serves: 6-8

1¼ pounds fresh ground turkey
2 tablespoons cooking oil
2 medium onions, chopped
1 cup chopped green bell pepper or celery
2½ cups chicken broth
1 (7 oz.) package elbow macaroni

1 (15 oz.) can tomato sauce
2 tablespoons vinegar
1½ teaspoons sugar
1 teaspoon chili powder
1 teaspoon garlic salt
6 tablespoons grated Parmesan cheese
1½ tablespoons snipped parsley

Heat oil in Dutch oven over medium high heat. Sauté ground turkey, onion, and pepper until meat is no longer pink. Drain, reserving juices. Set aside meat. Return juice to pan, add broth, and bring to a boil. Add uncooked macaroni; simmer 10 minutes, stirring often, until broth is almost absorbed. Stir in all remaining ingredients (except 2 tablespoons Parmesan cheese and parsley). Simmer 10 minutes. Transfer to serving dish; sprinkle with remaining Parmesan cheese and parsley.

TURKEY AND HAM TETRAZZINI

Serves: 6

1 (7 oz.) package spaghetti, cooked and drained
¼ cup slivered almonds, toasted
1 (10¾ oz.) can condensed cream of mushroom soup
1 (10¾ oz.) can condensed cream of chicken soup
¾ cup milk
2 tablespoons dry white wine

2 cups cut-up cooked turkey or chicken
½ cup cut up fully cooked smoked ham
½ cup chopped green bell pepper
½ cup halved pitted ripe olives
½ cup grated Parmesan cheese

Preheat oven to 350°.　　　　　　　　　　　　　2 quart casserole

Rinse cooked spaghetti with cold water to maintain firmness. Toast almonds in oven on ungreased pan for about 10 minutes, stirring occasionally. Continue until golden brown. Set aside. Mix condensed soups, milk, and wine in dish. Stir in spaghetti, turkey, ham, chopped pepper, and pitted olives. Sprinkle top of mixture with Parmesan cheese. Bake uncovered for 30 minutes or until hot and bubbly. Garnish with almonds.

SHRIMP CREOLE

Serves: 4

3 tablespoons butter
¼ cup diced green bell pepper
¼ cup minced onion
½ cup diced celery
1 tablespoon flour
1 (16 oz.) can peeled tomatoes
1 teaspoon salt
Dash of pepper

1 teaspoon sugar
1 bay leaf
1 sprig fresh parsley or ¼ teaspoon dried parsley flakes
1½ cups cooked shrimp
¼ teaspoon Worcestershire sauce
1½ cups steamed rice

Melt butter in saucepan. Add bell pepper, onion, and celery; sauté 5 minutes or until tender. Blend in flour. Gradually add tomatoes. Add salt, pepper, sugar, bay leaf, and parsley. Simmer gently, uncovered, 30 minutes. Remove bay leaf and parsley. Add shrimp and Worcestershire sauce; heat thoroughly. Serve over rice.

EASY GRILLED SHRIMP

Serves: 4-5

¼ cup olive oil
2 tablespoons chopped, fresh parsley
1 tablespoon minced garlic
Juice of one lemon
1 tablespoon ground black pepper

1½ pounds jumbo shrimp, peeled
15 slices lean bacon (or 5 slices Canadian bacon cut in strips)
½ cup diced white onion

13 x 9 baking pan

Mix oil, garlic, onion, lemon juice, pepper, and parsley together in pan. Marinate shrimp in oil mixture for 2 hours. Preheat grill to medium. Wrap bacon around each shrimp; secure with a toothpick. Grill shrimp for 5 minutes, turning and basting with marinade once. Remove shrimp from grill when opaque.

LEMON BARBECUED SHRIMP

Serves: 6

2½ pounds large fresh shrimp
½ cup lemon juice
¼ cup reduced-calorie Italian
    salad dressing
¼ cup water
¼ cup soy sauce

3 tablespoons minced fresh
    parsley
3 tablespoons minced onion
1-2 cloves garlic, crushed
½ teaspoon freshly ground
    pepper

13 x 9 baking pan

Peel and devein shrimp; place in dish. Combine remaining ingredients in a jar, cover tightly, and shake vigorously. Pour marinade over shrimp, cover, and refrigerate at least 4 hours. Thread shrimp onto skewers. Broil or grill 5 to 6 inches from medium heat for 3 to 4 minutes on each side, basting frequently with marinade. Serve immediately.

SHRIMP FETTUCCINE

Serves: 12

1½ cups butter
3 medium onions, chopped
3 ribs celery, chopped
2 green bell peppers, chopped
¼ cup flour (more may be
    needed to thicken)
4 tablespoons chopped parsley
3 pounds shrimp, cleaned and
    deveined
1 quart half-and-half

1 pound Velveeta cheese, cubed
    ½ inch
½ pound jalapeño cheese, cubed
    ½ inch
2 tablespoons chopped jalapeño
    peppers
3 cloves garlic, crushed
Salt and pepper to taste
1 pound fettuccine noodles
Parmesan cheese to taste

Preheat oven to 350°.

3 quart casserole dish

Lightly butter casserole dish. Melt butter in large saucepan. Add onions, celery, and bell peppers. Cook 10 minutes or until clear. Add flour; blend well. Cover and cook 15 minutes, stirring often. Add parsley and shrimp. Cover and cook 20 minutes, stirring frequently. Add cream, cheeses, jalapeños, and garlic. Mix well. Add salt and pepper to taste. Cook covered on low heat 20 minutes, stirring occasionally. Cook fettuccine according to package directions. Drain and add sauce. Mix thoroughly. Pour into casserole. Sprinkle with cheese. Bake for 12 minutes.

SHRIMP AND FRESH HERB LINGUINE
Serves: 4

12 ounces linguine
Salt to taste
7 tablespoons unsalted butter, divided
3 tablespoons olive oil
1 pound medium shrimp, peeled and deveined
8 cloves garlic, minced

⅓ cup chopped shallots
⅔ cup bottled clam juice
½ cup dry white wine
½ cup chopped fresh parsley
¼ cup chopped fresh dill or 1 tablespoon dried dillweed
1 teaspoon black pepper

Cook linguine in large pot of boiling salted water, stirring occasionally, until tender but firm (al dente). Drain pasta; return to pot to keep warm. Mix in 1 tablespoon unsalted butter. Heat olive oil in large heavy skillet over medium heat. Add shrimp; sauté, stirring frequently, until cooked through, about 3 minutes. Transfer shrimp to a bowl using a slotted spoon. Add garlic, shallots, and 1 tablespoon unsalted butter to skillet; cook 2 minutes. Add clam juice and dry white wine. Increase heat; boil until mixture is reduced by half, about 8 minutes. Reduce heat to medium-low; whisk in remaining 5 tablespoons unsalted butter. Return shrimp and accumulated juices to skillet. Add parsley, dill, black pepper, and pasta; toss to blend well. Season to taste with salt.

SHRIMP SCAMPI
Serves: 2

½ cup butter
2 teaspoons Worcestershire sauce
¼ cup sherry
1 clove garlic or equivalent of liquid garlic

2 tablespoons lemon juice
1 tablespoon sugar
1 pound raw shrimp, cleaned and deveined
¼ cup minced parsley
Parmesan cheese

Preheat broiler.   Shallow baking pan

Melt butter in a small saucepan over low heat. Add Worcestershire sauce, sherry, garlic, lemon juice, and sugar. Mix well. Arrange shrimp in a single layer in pan. Spoon sauce over shrimp. Broil for 8 minutes. Remove from broiler; let stand 15 minutes. Sprinkle parsley over shrimp; broil at high heat for 2 minutes. Serve over cooked rice; sprinkle with cheese.

SHRIMP PAESANO

Serves: 6-7

Shrimp:

1 pound jumbo shrimp, peeled
   and deveined
Flour

1 pint half-and-half
1 tablespoon malt vinegar
Oil

Preheat oven to broil.   13 x 9 baking pan

Marinate shrimp in cream for 10 minutes. Drain and dredge in flour. Over medium heat, sauté shrimp in oil and vinegar, without turning, for 3 minutes. Remove shrimp; place in baking pan in preheated oven. Broil 5 minutes.

Sauce:

Juice of 1 lemon
1 egg yolk
1-2 cloves garlic, minced

½ cup chopped chives
½ cup butter
½ cup chopped parsley

Mix egg yolk and lemon juice. Add half of butter; stir over low heat until melted. Add garlic and remaining butter. Stir briskly until butter melts and sauce thickens. Add parsley and chives. Place shrimp on serving plates; pour sauce over shrimp.

SHRIMP VEGETABLE STIR-FRY

Serves: 4

2 teaspoons cooking oil
1 pound medium shrimp,
   shelled and deveined
1 medium onion, sliced
1 medium green bell pepper,
   sliced into strips

3 cloves garlic, minced
¼ teaspoon salt
1 (14½ oz.) can diced
   tomatoes, pureed

Large heavy skillet or wok

Heat oil in large skillet or wok over medium-high heat. Stir-fry shrimp, onion, bell pepper, and garlic 6-8 minutes, or until shrimp turns pink and vegetables are crisp-tender. Add salt and tomatoes. Cook and stir 2-3 minutes, or until thoroughly heated. Serve over white rice.

SCAMPI ITALIANO

Serves: 4

1 pound (15-20) large shrimp, peeled and deveined
½ cup olive oil
1 teaspoon crushed, dried rosemary leaves
1 teaspoon Italian seasoning
1 teaspoon Cajun seasoning mix
2 tablespoons minced garlic

2 tablespoons capers
½ cup white wine
⅛ teaspoon Worcestershire sauce
½ cup chopped green onion, green tops only
½ cup shrimp or chicken stock
½ cup unsalted butter, softened

Warm serving plates

In large skillet, sauté shrimp over high heat, with oil, herbs, seasonings, garlic, and capers. Mix until shrimp are well coated and seared. Deglaze pan with wine, Worcestershire sauce, green onion, and stock. Continue cooking until mixture is reduced by half and shrimp are almost done. Turn heat to medium low, add softened butter, and shake pan vigorously to incorporate butter into sauce. Immediately arrange shrimp on warm serving plates; spoon sauce over each serving.

SPICY SHRIMP

Serves: 4

½ cup butter
1 tablespoon black pepper
2 ounces Worcestershire sauce
¼ teaspoon Tabasco sauce
¼ teaspoon garlic powder

2 teaspoons dried parsley
2 pounds medium shrimp, uncooked, peeled, and deveined
1 lemon, thinly sliced

Preheat oven to 400°.

13 x 9 baking pan

Melt butter in small saucepan; add black pepper, Worcestershire sauce, Tabasco, garlic powder, and dried parsley. Put shrimp into baking pan. Pour butter mixture over shrimp; top with lemon slices. Cover; bake 5-8 minutes, or until shrimp turn pink.

Add caraway seeds to boiling water to remove odor when cooking cabbage and shrimp. No caraway taste remains in the finished product.

SHRIMP SEBASTIAN

Serves: 4

½ cup whipping cream
1 tablespoon chopped garlic
2½ cups well-chilled unsalted butter, cut into tablespoon-size pieces
½ cup dry white wine
32 large shrimp, uncooked, peeled, and deveined

8 fresh or frozen artichoke hearts, cooked and quartered
2 cups thinly sliced mushrooms
2 cups thinly sliced green onion tops
Salt and ground black pepper to taste
8 French bread slices

Large heavy skillet

Boil cream and garlic in heavy medium saucepan until reduced by half, stirring occasionally, about 5 minutes. Remove from heat; whisk in 2 tablespoons butter. Set pan over low heat; whisk in remaining butter, 1 tablespoon at a time, removing from heat briefly if drops of melted butter appear. (If sauce breaks down at any time, remove from heat; whisk in 2 tablespoons butter). Keep sauce warm in water bath until ready to use; do not boil. Bring dry white wine to boil in large heavy skillet. Add shrimp and artichoke hearts; cook until shrimp turns pink, 3 to 5 minutes. Pour off all but 1 tablespoon wine. (If all wine is absorbed, add another tablespoon.) Add butter sauce, mushrooms, and green onion tops; bring to boil. Season with salt and pepper. Ladle into shallow bowls. Arrange 2 slices French bread atop each; serve.

SHRIMP VICTORIA

Serves: 4

1 pound raw shrimp, peeled
1 small onion, finely chopped
¼ cup butter
1 cup sliced fresh mushrooms
1 tablespoon flour
¼ teaspoon salt

Dash of pepper
½ cup sour cream
1 tablespoon vermouth
Juice of ½ lemon
Cooked rice

Sauté shrimp and onion in butter for 10 minutes or until shrimp are pink. Add mushrooms and cook 5 more minutes. Sprinkle in flour, salt, and pepper. Stir in sour cream, vermouth, and lemon juice; cook gently for 10 minutes. Be certain mixture does not boil. Serve over rice.

CRAB SMOTHERED SHRIMP

Serves: 4-6

2 dozen fresh jumbo shrimp, shelled and deveined
2 tablespoons butter
1 small onion, finely minced
½ cup finely minced green bell pepper
1 cup finely minced celery
1 tablespoon fresh parsley, chopped
1 pound fresh white crabmeat

1 teaspoon salt
1 teaspoon Worcestershire sauce
Dash Tabasco
½ cup seasoned bread crumbs
1 egg, beaten
½ cup melted butter (do not substitute)
2 cloves garlic, minced
Paprika

Preheat oven to 400°.

13 x 9 baking pan

Butterfly shrimp and spread flat in baking dish. Melt 2 tablespoons butter in large skillet and saute´ onion, bell pepper, and celery. Remove from heat and add parsley. Mix in crabmeat. Add remaining ingredients, except butter and toss well. Mound crab mixture over shrimp. Top with melted butter and sprinkle with paprika. Bake for 20 minutes or until shrimp is done.

JAMBALAYA

Serves: 12

2 cups uncooked white rice
2 pounds uncooked shrimp, deveined and peeled
½ cup butter or margarine, melted
1 (10 oz.) can French onion soup
1 (10 oz.) can beef broth

2 tablespoons Tony's Cajun seasoning
6-7 green onions, chopped
1 (15 oz.) can tomato sauce
1 (10 oz.) can mushrooms
1 green bell pepper, chopped
1 pound link sausage, sliced

Preheat oven to 350°.

3 quart casserole

Lightly grease casserole. Place rice on bottom of casserole. Mix all other ingredients; pour on top. Cook 1 hour and 15 minutes, uncovered.

SOFT SHELL BLUE CRABS

Serves: 4-6

8 large soft shell crabs
(prepared for cooking)

Marinade:
1 cup dry white wine
½ cup virgin olive oil
8-10 cloves garlic, finely minced
1 medium onion

2-3 bay leaves
½ teaspoon thyme
½ teaspoon rosemary
½ teaspoon allspice

Mix marinade ingredients well. Marinate the crab about 2 hours in the refrigerator. Grill over hot coals for 5 to 8 minutes.

FISH VERACRUZ

Serves: 8

1½ pounds orange roughy
1 cup sliced onions
½ cup sliced pimientos
4 cups chopped fresh tomatoes
1 teaspoon chopped green
chilies, seeded

4 teaspoons chopped capers
4 teaspoons chopped cilantro
Cilantro for garnish

Preheat oven to 350°.

13 x 9 baking pan

Wash fish with cold water and pat dry. Divide fish into eight 3-ounce portions. Arrange in baking dish. In a medium covered skillet, cook onions over very low heat, stirring frequently until soft. Add ¼ cup of chopped pimiento, tomatoes, chilies, capers, and cilantro. Cook until there is about 1 inch of juice in the skillet. Pour sauce over fish evenly and bake in preheated oven for 15 minutes, or until fish flakes easily with a fork (time will depend on the type and thickness of fish). Serve with sauce spooned over the top. Garnish with remaining pimiento and cilantro.

 Thaw fish in milk. The milk draws out the frozen taste and provides a fresh-caught flavor.

HERBED FISH

Serves: 4

4 (4-6 oz.) white fish fillets
   Garlic salt
¼ teaspoon dried oregano
¼ teaspoon dried tarragon
1 small garlic clove, minced
¼ teaspoon onion salt

¼ teaspoon dried parsley flakes
¼ teaspoon pepper
6 tablespoons butter or
   margarine, melted
3 tablespoons dry sherry

Sprinkle fillets with garlic salt. Sauté seasonings and garlic in butter in a large heavy skillet over medium heat for 1 minute. Reduce heat to low; add fish. Cook fish 4 minutes; turn and cook an additional 3-4 minutes, or until fish flakes easily when tested with a fork (do not overcook). Remove fish with a slotted turner; place on a heated platter, and keep warm. Pour sherry into skillet. Cook over high heat 30-45 seconds. Pour sherry mixture over fish. Serve immediately.

SOLE PICCATA

Serves: 4

4 fillets of sole, fluke, or
   flounder (1½ lbs.)
   Salt and pepper
½ cup flour
2 tablespoons oil

8 tablespoons butter, divided
2 tablespoons freshly squeezed
   lemon juice
2 tablespoons chopped fresh
   parsley

Large heavy skillet

Sprinkle fillets with salt and pepper. Dredge in flour; shake off excess. Heat 2 tablespoons of oil and 2 tablespoons butter in skillet over medium-high heat until hot but not smoking. Add fillets; cook, turning once, until crisp and golden, about 4 minutes. Transfer fillets to warm plates. Reduce heat to low; add remaining 6 tablespoons butter to skillet, cook until butter bubbles, about 1 minute. Remove skillet from heat; stir in lemon juice and parsley. Pour butter over fillets; serve immediately.

If desired you may use 2 skillets to expedite the cooking; sauce can still be made in one pan.

 Cook frozen fish while it is still a bit icy in the middle. Fish takes only a few minutes longer to cook and stays moist and juicy.

ORANGE ROUGHY WITH RED PEPPERS

Serves: 4

1 pound orange roughy or lean fish fillets
1 teaspoon olive or vegetable oil
1 small onion, cut into thin slices
2 red or green bell peppers, cut into julienne strips

1 tablespoon snipped fresh thyme leaves or 1 teaspoon dried thyme leaves
¼ teaspoon black pepper

If fish fillets are large, cut into 4 serving pieces. Heat oil in 10 inch nonstick skillet. Layer onion and bell peppers in skillet; sprinkle with half of the thyme and half of the black pepper. Place fish over bell peppers and sprinkle with remaining thyme and black pepper.

Cover and cook over low heat 15 minutes. Uncover and cook until fish flakes easily with fork, 10 to 15 minutes longer.

Microwave directions:

Omit oil. Layer onion and bell peppers in a rectangular microwave safe dish, sprinkle with half of the thyme and half of the black pepper. Cover with vented plastic wrap and microwave on high (100%) 2 minutes. Arrange fish, thickest parts to outside edges, on bell peppers; sprinkle with remaining thyme and black pepper. Cover with vented plastic wrap and microwave 4 minutes; rotate dish ½ turn. Microwave until fish flakes easily with fork, 3 to 5 minutes longer. Let stand covered 3 minutes.

For extra-crisp bell peppers when cooking this dish in the microwave, omit the step of cooking the bell peppers before adding the fish.

BAKED SALMON PATTIES

Serves: 8

1 (14¾ oz.) can red salmon
1 medium onion, grated
1 egg, beaten
½ cup milk
½ cup mayonnaise

1½ cups crushed corn flakes or almost any unsweetened cereal, divided
½ cup margarine, melted

Preheat oven to 350°.   15 x 10 jelly roll pan

Flake salmon, including juice and bones. Stir in onion, egg, milk, mayonnaise, and 1 cup of the crushed cereal. Form into 10 patties; roll patties in remaining ½ cup of crushed cereal to coat. Line jelly roll pan with foil; spray foil with non-stick spray. Place patties in pan; pour melted margarine over patties, bake for 30 minutes.

GRILLED SALMON WITH LEMON BUTTER

Serves: 4

2 tablespoons olive oil
1 teaspoon dried thyme (or 1 tablespoon fresh thyme)
2½ tablespoons butter, room temperature

½ teaspoon lemon juice
1 teaspoon grated lemon zest
4 (1-inch thick) salmon steaks
Salt and pepper

Preheat grill to medium.

Combine oil and thyme. Mix butter with lemon zest and juice. Put butter mixture on a piece of plastic wrap; roll to form a 1 inch cylinder. Put in the refrigerator (or in freezer if in a hurry) to firm while the fish is cooking. Brush fish with thyme and oil; sprinkle with salt and pepper. Cook fish on medium grill, brushing occasionally with oil until opaque (4-5 minutes per side). Top each hot steak with 2 thin slices of lemon butter; serve.

SALMON LOAF

Serves: 8-10

1 (14¾ oz.) can pink salmon
¼ cup mayonnaise-type salad dressing
1 (10¾ oz.) can cream of celery soup
1 egg

½ cup chopped onion
¼ cup chopped green bell pepper
1 tablespoon fresh or bottled lemon juice
1 cup bread crumbs

Preheat oven to 350°.   9 x 5 loaf pan

Grease loaf pan. Drain salmon; remove bones and skin. Combine salmon with salad dressing, soup, egg, onion, pepper, lemon juice, and bread crumbs. Bake 1 hour.

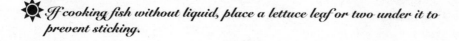

If cooking fish without liquid, place a lettuce leaf or two under it to prevent sticking.

SALMON STEAKS WITH PISTACHIO BUTTER Serves: 6

Pistachio Butter:

¾ cup unsalted butter or margarine, softened

3 tablespoons minced shelled pistachio nuts

½ teaspoon grated lime zest

Soften the butter in a bowl with the back of a wooden spoon or in a food processor fitted with the steel blade. Mix in the nuts and lime zest. Spoon the butter into a crock, cover, and chill until needed. Remove the butter from the refrigerator 45 minutes before you use it.

Fish:

6 (6-7 oz.) salmon steaks

¼ cup extra-virgin olive oil

¼ cup freshly squeezed lime juice

1 teaspoon dried basil

2 limes, sliced, for garnish

Preheat broiler.

Line shallow broiler pan with aluminum foil. Arrange the salmon steaks in the pan. Mix together the oil, lime juice, and basil. Brush the top surface of the salmon steaks with the oil mixture. Broil the salmon steaks 4 to 6 inches from heat source for 4-5 minutes. Turn the salmon steaks over with a spatula. Brush the fish with the oil mixture and continue broiling until the fish flakes easily when tested. Using a spatula, transfer the salmon steaks to individual plates. Garnish with lime slices and a dollop of Pistachio Butter on top of each steak. Serve immediately.

LINGUINE WITH BROCCOLI AND SCALLOPS Serves: 4

4 cups broccoli flowerets

½ pound linguine

12 tablespoons butter, divided

2 teaspoons minced garlic

1 pound bay scallops

2 teaspoons salt

1 teaspoon pepper

¾ cup grated Parmesan cheese

Blanch broccoli in boiling salted water; drain; set aside. Cook pasta 8 to 10 minutes and drain. Toss with 4 tablespoons butter. Heat remaining butter; add garlic; sauté until soft. Add scallops, salt, and pepper; cook until scallops are opaque. Add broccoli, heat thoroughly. Toss with cheese. Serve.

SEAFOOD CRÊPES
Serves: 24

Seafood:

¼ cup butter, divided
1½ pounds scallops
2 pounds shrimp

1 bunch scallion tops, chopped
1 pound crab meat, uncooked

Sauté scallops in 2 tablespoons butter; reserve liquid; set scallops aside. Sauté shrimp and scallion tops in 2 tablespoons butter, reserve liquid. Combine scallops, shrimp, and crab. Combine seafood liquids and reserve for roux.

Roux:

½ cup butter
⅔ cup flour
6 cups warm milk
Reserved seafood liquid

⅓ cup dry sherry
2 cups grated Parmesan cheese
3 cups grated Gruyère cheese

Melt butter in 4 quart pot; add flour; constantly stir. Add milk in a steady stream; whisk until mixture thickens; add seafood liquid; stir. Add sherry, cheeses, and seafood. Cook 10-15 minutes over low heat until mixture is thick and creamy, stirring frequently to avoid burning. May be prepared a day ahead.

Basic Crêpes:

1 cup flour
½ cup milk at room temperature
3 large eggs

2 tablespoons unsalted butter,
  softened
½ teaspoon salt

Food processor.   6-8 inch Teflon pan

Combine all ingredients and mix in food processor on medium high for 3 minutes. Cover and let stand 20 minutes. Meanwhile, heat pan on medium low heat, spray with nonstick and remove from heat. Pour ⅛ cup, (or less) batter into pan; tilt pan around; return to heat; cook until crêpe is lightly browned. Flip crêpe and cook for a minute or less; remove from heat. Crêpes may be prepared 3 to 4 days ahead; covered and refrigerated.

Seafood Crêpes:

Preheat oven to 350°.   4 (13 x 9) baking pans

Lightly grease baking pans. Fill crêpes with 3 tablespoons seafood mixture; place seam side down in pan. Bake for 25-30 minutes until thoroughly heated. Crêpes may be filled and refrigerated the day before serving.

Crêpe batter needs to be made 3 times in order to serve 24.

SEAFOOD WILD RICE CASSEROLE

Serves: 10-12

1 (6 oz.) package brown and
 wild rice mix
1 pound fresh or frozen crab
 meat
1 pound shrimp, cooked and
 peeled
3 (10¾ oz.) cans mushroom
 soup

1 cup chopped celery
1 (4 oz.) jar chopped pimientos,
 drained
1 (4½ oz.) can mushrooms,
 drained
⅓ cup chopped onion
1 cup chopped green bell
 pepper

Preheat oven to 325°. 4 quart casserole

Lightly grease casserole. Prepare rice mix according to package directions. Add remaining ingredients, stirring well. Spoon mixture into prepared casserole. Bake for 1 hour.

VEAL BRISKET POCKET ROAST

Serves: 6-8

5-6 pound veal brisket pocket
 roast

Garlic powder to taste
Paprika to taste

Preheat oven to 325°. Roasting pan

Season roast, set aside while making the following stuffing.

Stuffing:

1 onion, chopped
3 tablespoons butter
1 (4 oz.) can mushroom pieces,
 drained
¼ cup diced green bell pepper
¼ cup diced celery

¼ cup parsley
2 eggs, beaten
¼ cup chicken broth
¾ cup herb stuffing mix
¾ cup cornmeal mix

Sauté onion in butter; add mushroom pieces, green pepper, celery, and parsley. Combine with stuffing mixes; add eggs and chicken broth. Cool stuffing mixture; stuff seasoned roast. Place roast over a layer of the following ingredients:

2 medium onions, sliced
6-8 stalks celery

¼ cup water

VEAL PICATTA
Serves: 4

1 ½ pounds veal round or sirloin,
cut ¼-½ inch thick
Salt and pepper to taste
Flour to dust
3 tablespoons butter
1 tablespoon olive oil
2 cloves garlic, minced
½ pound fresh mushrooms,
sliced

2 tablespoons fresh lemon juice
½ cup dry white wine
2 teaspoons capers plus 1
teaspoon caper juice
(optional)
3 tablespoons minced parsley
½ lemon, thinly sliced

Sprinkle veal with salt and pepper on both sides; dust lightly with flour. Heat butter and olive oil in a large skillet. Add veal; brown on both sides. Remove veal from skillet. Add garlic and mushrooms to pan; cook one minute. Return veal to pan. Add lemon juice and white wine; cover and simmer for 20 minutes or until veal is tender. Add capers. Remove to a warm platter. Sprinkle with parsley; garnish with lemon slices.

VEAL MEDALLIONS WITH SHALLOTS AND APPLES
Serves: 4

1 pound veal scallops
¾ cup fresh grated Parmesan
cheese
¼ cup flour
Seasoned salt and pepper
4 tablespoons olive oil or butter

2 green apples, peeled, cored,
and sliced
4 shallots, cut in wedges or 1
onion cut in wedges and
halved
1 cup chicken broth

Mix cheese and flour. Salt and pepper veal and dredge in the cheese/flour mixture. Heat oil or butter in large skillet over medium heat. Cook veal for 3 minutes on each side or until brown. Remove veal from pan. Keep warm. Add apples and shallots to pan. Sauté 3 minutes; add chicken broth and simmer for 5 minutes. Put veal into broth and heat 1 minute longer. Pour into casserole dish and serve.

Side Dishes

Refer to Mexican Cuisine section for additional recipes in this category.

Don't Wait to Vaccinate

Our public awareness campaign, promoted by our Public Relations Committee in cooperation with the Hidalgo County Health Department, has been held during the National Immunization Awareness Week. Posters and handouts, along with radio PSA's and newspaper coverage in both English and Spanish, have been distributed to various targeted areas. Within the first year, over 400 at risk children were vaccinated. Our goal has been to educate adults that many childhood diseases can be prevented through vaccination.

 These recipes are pictured on the previous page.

BURGUNDY APPLES

Serves: 12

12 Winesap apples, peeled, cored, and quartered
2 cups sugar
2 sticks cinnamon

¼ cup Burgundy wine
¼ cup tarragon vinegar
Red food coloring

In a large pan combine sugar, cinnamon, wine, and vinegar. Bring to a boil, stirring until sugar is melted. Add food coloring until a dark red color is achieved. Reduce heat. Add apples, a few at a time; cook until tender. Remove and add more apples, repeat until all are cooked. Add more food coloring if necessary. Serve warm or cold.

May substitute Red Delicious apples.

ARTICHOKE AND MUSHROOM BAKE

Serves: 6

½ pound thinly sliced fresh mushrooms
2 tablespoons butter or margarine
2 (10¾ oz.) cans cream of mushroom soup
½ cup milk

2 tablespoons dry sherry
2 tablespoons Worcestershire sauce
2 (14 oz.) cans drained artichoke hearts
¼ cup grated Parmesan cheese

Preheat oven to 375°.

12 x 7 casserole dish

Lightly grease casserole dish. In a skillet, sauté mushrooms in butter until soft. Add remaining ingredients, except artichokes and cheese. Salt and pepper to taste and heat. Place artichokes in the baking dish and pour sauce over them. Sprinkle with cheese and bake uncovered for 20 minutes.

 If you scorch the inside of a saucepan, fill the pan halfway with water and add 1/4 cup baking soda. Boil a while until the burned portions loosen and float to the top.

STUFFED ARTICHOKES

Serves: 8

16 artichokes	⅛ teaspoon pepper
Lemon slices	⅛ teaspoon nutmeg
1 clove garlic	¾ cup milk
4 tablespoons butter	1½ tablespoons lemon juice
2 tablespoons flour	2 egg yolks
⅛ teaspoon salt	

Preheat oven to 325°.   13 x 9 baking pan

Butter baking pan. Cook artichokes in boiling water with lemon slices and garlic. Invert and cool until they can be handled. Before beginning sauce, remove leaves from each artichoke. Using a spoon, scrape pulp from inner sides of leaves as they are removed. Cut out the tough chokes from each vegetable and discard along with scraped leaves. Carefully scoop out heart but leave enough to make a container for the stuffing. Chop up the artichoke bottoms, and add to the pulp from inner sides of leaves, mashing until smooth. Melt the butter. Add flour and let the roux cook 2-3 minutes without browning. Add seasonings and hot milk; cook until thick. Add lemon juice. Beat egg yolks and add a little of the hot mixture to them. Then pour into sauce. Add sauce to the pulp, mixing well. Mound mixture into artichoke hearts. Place in pan and bake for 25 minutes.

This takes time. Plan to prepare 1-2 days in advance.

ASPARAGUS CONFETTI

Serves: 8

¾ cup oil	1 tablespoon sugar
½ cup vinegar	4 tablespoons sweet pickle relish
4 teaspoons chopped, drained pimientos	3 tablespoons chopped parsley
2 teaspoons salt	4 (15 oz.) cans asparagus
3 tablespoons chopped green onions	

Combine all ingredients except asparagus, put in a jar and refrigerate. Chill asparagus in cans. Drain, then layer asparagus in serving dish. Pour mixture over asparagus and serve.

SOUR CREAM ASPARAGUS

Serves: 6-8

2 (10½ oz.) cans asparagus
  tips, well drained
1 cup sour cream

5 tablespoons mayonnaise
5 tablespoons grated Parmesan
  cheese

Set oven to broil.   13 x 9 baking pan

Arrange asparagus in baking pan; set aside. Combine remaining ingredients, mix well, and spread over asparagus. Broil until lightly browned and bubbly (do not overcook). Serve immediately.

Sauce may be served over broccoli, cauliflower, or any vegetable.

BAKED BEANS

Serves: 4-6

2 slices bacon
4 slices onion
½ green bell pepper, chopped
1 (16 oz.) can pork and beans
2 tablespoons ketchup

2 tablespoons brown sugar
1 teaspoon Worcestershire
  sauce
½ teaspoon prepared mustard

Preheat oven to 350°.   9 x 9 baking pan

Lightly grease baking pan. Cook bacon until crisp. Drain and crumble. Sauté onion and bell pepper in bacon grease. Drain. Combine onion mixture with bacon, beans, and other ingredients in baking pan. Cover and bake for 45 minutes.

BLACK-EYED PEAS

Serves: 6

1 (1 lb.) package dried black-
  eyed peas
1 can beer
½ onion, chopped
½ pound ham, chopped

1 teaspoon dry mustard
1 (48 oz.) can spiced tomato
  juice, or V-8 juice
1 green bell pepper, chopped

Soak peas overnight in beer. Combine the onion, ham, dry mustard, bell pepper, and juice in pot; cook until peas are tender, about 2 hours. If more liquid is needed, add half beer and half water.

BROCCOLI PASTA CASSEROLE

Serves: 6-8

2 bunches fresh broccoli, cut into small pieces

1 (12 oz.) box rotini pasta noodles

2 (10¾ oz.) cans of cream of mushroom soup or cream of celery soup

1 clove garlic, minced

1 can onion rings (optional)

Preheat oven to 325°.   13 x 9 casserole dish

Lightly grease casserole dish. Cook broccoli pieces, covered in boiling salted water about 5 minutes. Drain. Meanwhile, cook pasta according to package directions. Drain. Combine broccoli, pasta, soup, and garlic. Pour into casserole dish. Top with onion rings if desired. Bake uncovered for 30 minutes.

AUSTRALIAN CABBAGE

Serves: 6

½ head cabbage, shredded

½ teaspoon salt

2 teaspoons sugar

2 tablespoons butter, melted

1 bacon slice, diced

1 white onion, diced

2 hard boiled eggs, chopped

2 tablespoons grated Cheddar cheese

Preheat oven to 400°.   2 quart casserole dish

Lightly grease casserole dish. Combine cabbage, salt, and sugar. Cook in saucepan, occasionally stirring, for 10 minutes until crisp. Meanwhile, in skillet, sauté bacon and onion in butter until the onion is soft. Combine with cabbage and drain mixture. Add eggs to cabbage. Pour mixture into casserole dish and cover with grated cheese. Bake uncovered for 20 minutes.

No more tears when peeling onions if you place them in the freezer for 4 or 5 minutes first.

SKILLET CREOLE CABBAGE

Serves: 8

2 tablespoons margarine
½ cup chopped onion
1 cup chopped green bell
  pepper
1 cup diced celery

2 cups diced fresh tomatoes
4 cups shredded cabbage
1½ teaspoons salt
1 teaspoon sugar
¼ teaspoon pepper

Melt margarine in heavy skillet. Add remaining ingredients. Cook over medium heat 10-12 minutes or until vegetables are tender. Do not overcook.

CARROTS AND GRAPES

Serves: 6

3 cups sliced cooked carrots

2 cups green seedless grapes

Glaze:

⅓ cup brown sugar, firmly
  packed
¼ cup butter

¼ teaspoon cinnamon
⅛ teaspoon nutmeg
1 teaspoon lemon juice

Combine carrots and grapes in a bowl. To prepare glaze, combine sugar, butter, cinnamon, nutmeg, and lemon juice in saucepan. Heat until blended and hot. Pour hot glaze over carrots and grapes. Stir gently to coat. Serve immediately.

CRANBERRY CARROTS

Serves: 4

5 medium fresh carrots, peeled
¼ cup butter
½ cup canned cranberry jelly
  sauce

4 tablespoons brown sugar
½ teaspoon salt

Slice carrots crosswise or on the bias about ½ inch thick. Cook in saucepan filled with salted water 6 to 10 minutes (just until carrots are tender). In a medium size saucepan, combine butter, cranberry jelly sauce, brown sugar, and salt. Cook on low heat until mixture has melted. Drain carrots and add to cranberry mixture. Return to heat and cook approximately 5 minutes or until carrots are glazed.

SUNSHINE CARROTS

Serves: 6

7-8 medium carrots
1 tablespoon brown sugar
1 teaspoon cornstarch
¼ teaspoon ground ginger
¼ teaspoon salt

¼ cup orange juice
2 tablespoons butter
½ teaspoon parsley or
½ teaspoon grated lemon rind
(optional)

Bias-slice carrots crosswise about ½ inch thick. Cook in saucepan containing 1 inch boiling water until tender, about 10-15 minutes. Drain. In another saucepan, combine sugar, cornstarch, ginger, and salt. Add orange juice. Cook, stirring constantly until thickened and bubbly. Boil 1 minute more. Remove from heat. Stir in butter. Pour over hot carrots, tossing to coat evenly. Garnish with parsley or lemon rind.

CORN PUDDING

Serves: 4-6

½ green bell pepper, finely
chopped
1 small onion, finely chopped
2 tablespoons bacon drippings
or olive oil
1 tablespoon flour

1 egg, beaten
1 cup milk
1 teaspoon salt
1 (16 oz.) can creamed corn
1 cup bread crumbs, divided

Preheat oven to 350°.   13 x 9 casserole dish

Lightly grease casserole dish. Sauté bell pepper and onion in bacon drippings. Add 1 tablespoon flour and mix. Add and mix well egg, milk, salt, corn, and ½ cup bread crumbs. Pour mixture into pan. Sprinkle remaining bread crumbs on top. Bake 30 minutes or until browned and set.

 The darker the orange color of carrots; the greater the content of Vitamin A.

SOUTHERN CORN PUDDING

Serves: 10-12

3½ cups milk
1 cup yellow cornmeal
3 tablespoons + 1 teaspoon sugar
1 teaspoon salt
3 large eggs

½ teaspoon vanilla extract
1 teaspoon baking powder
1 (8 oz.) can whole kernel corn, undrained
4 tablespoons butter, softened

Preheat oven to 300°.   11 x 7 baking pan

In a medium saucepan, heat milk to boiling. Then, stirring constantly with a wire whisk, mix in cornmeal, sugar, and salt. Reduce heat so that mixture simmers, and cook, stirring constantly, for 5 minutes. Remove from heat. Beat eggs well and add to mixture. Then add vanilla, baking powder, undrained corn, and 2 tablespoons of butter. Mix thoroughly, pour into baking pan, dot top with remaining butter, and bake for 20 minutes or until top is lightly browned. Serve hot!

WATER CHESTNUT AND MUSHROOM DRESSING

Serves: 10-12

Finely chopped giblets
2 cups water
1 cup finely chopped celery
1 cup finely chopped green bell pepper
1 cup finely chopped onion
1 (5 oz.) can sliced or chopped water chestnuts

1 (8 oz.) can chopped mushrooms
1 cup butter or margarine
1-1½ cups giblet broth
1 package cornbread stuffing
1 package plain bread stuffing

Preheat oven to 350°.   13 x 9 baking pan

Simmer giblets in about 2 cups water to make giblet broth. In a big bowl mix together celery, bell pepper, onion, water chestnuts, mushrooms, and giblets. In sauce pan, melt butter in broth; stir in and lightly toss stuffings. Mix with chopped vegetables. Stuff turkey or bake separately. Add broth to desired moistness.

EGGPLANT AND CHEESE CASSEROLE

Serves: 4-6

1 large eggplant
1 green bell pepper, chopped
1 clove garlic, minced
1 onion, chopped
2 celery ribs, chopped
2-3 tablespoons olive oil

2 eggs, well beaten
1 cup bread crumbs
1 (4 oz.) can mushrooms, drained
Salt and pepper to taste
1½ cups grated Cheddar cheese

Preheat oven to 350°.   1½ quart casserole dish

Lightly grease casserole dish. Peel and dice eggplant into 2 inch cubes. Boil in salted water about 15 minutes. Drain well and mash. Sauté bell pepper, garlic, onion, and celery in olive oil. Add sautéed vegetables to mashed eggplant. Add eggs, bread crumbs, salt, pepper, and mushrooms. Place in casserole; bake 1 hour uncovered. During last 15 minutes of baking, top with grated cheese. Continue to bake until cheese is melted and bubbly.

FRIED EGGPLANT

Serves: 4-6

1 large eggplant

Milk batter:
2 eggs, beaten
⅔ cup milk

1 tablespoon oil

In a bowl, combine ingredients; set aside.

Flour mixture:
1 cup flour
1 teaspoon baking powder
2 teaspoons Italian seasoning

1 teaspoon onion powder
⅛ teaspoon red pepper
⅛ teaspoon garlic powder

In a bowl, combine ingredients; set aside.

Vegetable oil

Peel eggplant and cut into strips. Soak in salted water for 1 hour; drain. Meanwhile prepare milk batter and flour mixture. Dip eggplant in milk batter and then in flour mixture. Heat oil in skillet. Fry eggplant in oil until golden brown.

HOT FRUIT COMPOTE

Serves: 10-12

1 (29 oz.) can sliced peaches
1 (29 oz.) can pineapple chunks
1 (29 oz.) can sliced pears
1 (29 oz.) can mandarin oranges
1 (29 oz.) can sliced mangos
1 (29 oz.) can mixed fruit for salads
1 (15 oz.) can Bing cherries or maraschino cherries
½ cup raisins

½ cup brown sugar, firmly packed
½ cup amaretto or sherry
1-2 teaspoons ginger (to taste)
¼ teaspoon curry powder (to taste)
¼ teaspoon nutmeg (to taste)
1-2 teaspoons cinnamon (to taste)
4 tablespoons butter
3-4 bananas, sliced

Preheat oven to 350°.   3 quart casserole dish

Drain all canned fruits; add raisins. Mix well with brown sugar, amaretto or sherry, ginger, curry, nutmeg, and cinnamon. Dot with butter; bake uncovered for 1 hour 15 minutes. Add bananas; bake for 15 minutes more. Keep basting during baking.

GREEN BEANS AND PEARS

Serves: 8

3 pears
¼ pound bacon, fried crisp and crumbled
1 tablespoon bacon drippings
1 tablespoon butter
1 tablespoon flour

¼ cup white wine
1 cup chicken broth
2 pounds fresh green beans, blanched
1 teaspoon salt
1 teaspoon white pepper

Peel and dice pears into 2 inch cubes. Sauté pears in bacon drippings and butter, until tender; remove. Stirring in flour, cook for 2 minutes. Add wine and chicken broth; simmer until thick. Add green beans and seasonings; cook until beans are tender. Stir in pears and simmer for 1 minute. Serve immediately, topped with crumbled bacon.

 To remove hard water stains from your stainless steel pans, soak a cloth with rubbing alcohol and apply to the area. Wash with hot water before using again.

SWISS BEANS

Serves: 10

2 tablespoons butter
2 tablespoons flour
½ teaspoon salt
⅛ teaspoon pepper
1 tablespoon sugar

1 small onion, grated
1 cup sour cream
2 (16 oz.) cans French cut
green beans, drained
Swiss cheese, grated

Preheat oven to 350°.

8 x 8 casserole dish

Grease casserole dish. Melt butter and stir in flour, salt, pepper, sugar, and onion; gradually add sour cream. Cook, stirring until thickened. Mix in beans, put in casserole, and sprinkle with cheese. Heat until cheese melts and browns slightly.

JALAPEÑO PIE

Serves: 8

3 eggs
2 cups heavy cream
¼ teaspoon salt
¼ teaspoon black pepper
¼ teaspoon nutmeg
¾ cup chopped, seeded jalapeño peppers (or less, according to taste)

¾ cup cooked and crumbled bacon
1½ cups Gruyère cheese
1 (9 inch) deep dish pie shell, baked

Preheat oven to 375°.

Mix eggs, heavy cream, salt, pepper, and nutmeg. Place peppers and bacon in bottom of pie shell. Add cheese, then egg mixture. Bake for 40 minutes. Cool for 30 minutes, cut and serve. (May be prepared as much as 8 hours in advance.)

 Lumpless gravy can be your triumph if you add a pinch of salt to the flour before mixing it with water.

SPICY MACARONI AND CHEESE

Serves: 8

1 (12 oz.) package large elbow macaroni
6 slices thick cut bacon, cut crosswise in 1 inch pieces
3-4 stalks celery, chopped
1 small or medium onion, chopped

1 medium green bell pepper, chopped
1 (10 oz.) can diced tomatoes and green chilies, with juice
1½ pounds Velveeta cheese

Preheat oven to 350°.   2 quart casserole dish

Lightly grease casserole dish. Boil macaroni and drain. In a large frying pan, fry bacon until crisp; remove from pan. Sauté celery, onion, and bell pepper in bacon grease over medium heat, about 10 minutes. Add tomato mixture; simmer over low heat about 5 minutes. Cube 1 pound Velveeta cheese; mix with macaroni. Add vegetable mixture and bacon to macaroni and cheese mixture. Mix well; place in dish. Slice remaining ½ pound cheese; place on top. Cover with foil; bake until all cheese is melted.

For less fat use Velveeta Light or omit the sliced cheese for topping.

NOODLE KUGEL

Serves: 24

1 (16 oz.) package wide egg noodles
4 eggs
½ cup granulated sugar
6 tablespoons butter or margarine, melted, divided
1 quart buttermilk

1 teaspoon vanilla extract
1 teaspoon cinnamon
½ cup white raisins
¾ cup brown sugar, firmly packed
1 cup corn flake crumbs

Preheat oven to 350°.   3 quart casserole dish

Grease casserole dish. Cook noodles according to package directions. While noodles are cooking, beat together the eggs, sugar, 4 tablespoons butter or margarine, buttermilk, vanilla, and cinnamon. Add raisins and drained noodles. Pour into prepared dish and bake for 45 minutes. While mixture is baking, blend brown sugar, corn flake crumbs, and remaining butter or margarine. Spread over baked noodles and cook an additional 30 minutes. May be frozen.

PENNE WITH BASIL AND MOZZARELLA
Serves: 4

2 ounces mozzarella cheese
(about ½ cup)
1 ounce Parmesan cheese
(about ¼ cup grated)
½ pound Penne or other pasta

2½ tablespoons butter
2 tablespoons olive oil
¾ teaspoon salt
¼ teaspoon pepper
¼ cup chopped fresh basil

Cut the mozzarella into very thin slices, then into 1-inch squares. Grate the Parmesan. Cook the Penne in boiling salted water; drain, and return to pot. Add the cheese, butter, oil, salt, and pepper. Toss over low heat until mozzarella starts to melt. Taste and add additional salt, and pepper, if needed. Serve immediately; sprinkle with basil.

BAKED VALLEY ONIONS
Serves: 1

1 large onion
1 whole bay leaf
1 teaspoon seasoned salt
1 teaspoon seasoned pepper

Dash garlic powder
1 tablespoon butter
2 tablespoons Worcestershire
sauce

Preheat oven to 350°.   Cookie sheet

Cut off ends of onion; remove outer layer or two. Place core end down, cut center out (about ½ to ¾ inch), and quarter onion without cutting completely to bottom, (onion should still be fairly intact). Place on square of foil large enough to wrap around onion. Slide bay leaf down in cut of onion. Sprinkle top with salt, pepper, and garlic powder. Put butter in center pocket; pour Worcestershire sauce over top. Seal foil tightly; bake about 1½ hours, or until transparent.

When in season, 1015's are the best.

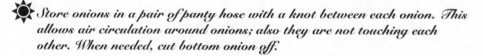

Store onions in a pair of panty hose with a knot between each onion. This allows air circulation around onions; also they are not touching each other. When needed, cut bottom onion off.

NUTTY ONIONS
Serves: 8-10

3 large onions
1 teaspoon freshly ground
 pepper, divided
1 (8 oz.) package mozzarella
 cheese, shredded
¼ cup chopped pecans or
 walnuts

1 cup heavy cream
1 (9 inch) pie crust, unbaked
1 egg yolk
1 tablespoon water

Preheat oven to 350°.   3 quart round casserole dish

Slice onions into ¼-inch thick rings. Steam onions, rings intact, for 5 minutes. Place ½ of the steamed onions in the bottom of casserole dish keeping rings intact. Sprinkle ½ teaspoon pepper over onions. Layer ½ of the cheese over onions, followed by the nuts. Layer remaining onions and then the cheese. Sprinkle ½ teaspoon pepper over layers followed by the heavy cream. Cover top of the dish with pie crust forming a rim around the edge. Just before baking, beat egg yolk and water together and brush over pie crust. Bake for 30-40 minutes.

CILANTRO SCALLOPED POTATOES
Serves: 8

¼ cup diced onion
¼ cup celery leaves
2 sprigs cilantro
3 tablespoons flour
¼ cup butter or margarine
1½ teaspoons salt
¼ teaspoon pepper

1½ cups milk
6 medium potatoes, peeled,
 cooked, and sliced
1-2 cups grated sharp Cheddar
 cheese
Paprika

Preheat oven to 350°.   13 x 9 baking pan

Slightly butter pan. Blend onion, celery, cilantro, flour, butter, salt, pepper, and milk in blender, mixing thoroughly. (This sauce may be made the day before.) Arrange potato slices in pan. Pour mixture over potatoes; sprinkle with grated cheese and paprika. Bake about 50 minutes, until bubbly and brown.

ONION TART

Serves: 8

Crust:

1¼ cups cracker crumbs

4 tablespoons butter, softened

With fingers, mix cracker crumbs with butter. Press into bottom and sides of pie plate.

Filling:

4 cups thinly sliced onions
6 tablespoons butter, divided
4 tablespoons flour
1 cup hot milk
½ cup hot chicken stock

½ cup sour cream
1 egg yolk, beaten
Salt and pepper
1-1½ cups shredded Longhorn cheese

Preheat oven to 350°.   9 inch pie pan

Sauté onions in 2 tablespoons butter in a heavy-bottomed pan until transparent. Melt 4 tablespoons butter, stir in flour and cook 1 minute. Remove from heat; stir in hot milk and chicken stock. Return to heat and stir until thickened. Add sour cream mixed with the egg yolk. Season to taste with salt and pepper. Mix sauce with onions and pour into crust. Sprinkle cheese over top and bake for 25-30 minutes.

JALAPEÑO POTATOES

Serves: 6-8

5 medium potatoes
1 small green bell pepper, slivered
1 small jar pimientos
Salt and pepper to taste
4 tablespoons butter or margarine
1 tablespoon flour

1 cup milk
½ roll garlic cheese, cubed or grated
½ roll jalapeño cheese, cubed or grated
Optional: 1 roll of either cheese may be used instead of ½ of each

Preheat oven to 350°.   3 quart casserole dish

Lightly butter casserole dish. Boil unpeeled potatoes in salted water until tender, but not falling apart. When cool enough to handle, peel and slice. Layer in dish with slivered bell pepper and pimiento. Salt and pepper each layer. Melt butter or margarine in saucepan; add flour and stir until well blended. Gradually add milk, stirring constantly. Add cheeses and melt. Pour over the potatoes and bake 45 minutes to 1 hour.

POTATO CASSEROLE

Serves: 4

4 large baking potatoes, peeled and thinly sliced
8 tablespoons butter or margarine, sliced into pats
1 onion, thinly sliced
1 (10¾ oz.) can cream of mushroom soup

1 (8 oz.) box fresh sliced mushrooms
Salt and pepper to taste
½ cup milk

Preheat oven to 350°.   3 quart casserole dish

Lightly grease casserole dish. Arrange a layer of potatoes topped with butter or margarine pats, onions, dots of soup, and mushrooms in casserole. Sprinkle with salt and pepper. Continue layering until all ingredients except milk are used. Pour milk over layers. Bake for 1½ hours.

If top browns too much, cover with foil and continue baking.

RANCH POTATOES

Serves: 4

8 small new potatoes, quartered
1 tablespoon butter or margarine
1 tablespoon chives

3 tablespoons Hidden Valley Ranch bottled salad dressing
Salt to taste

2 quart microwaveable casserole dish

Microwave potatoes 4-5 minutes until soft, but not mushy. Stir in butter and chives until potatoes are coated. Microwave 2 more minutes. Just before serving, coat with dressing. Salt to taste.

 A dampened paper towel or terry cloth brushed downward on a cob of corn will remove every strand of corn silk.

TWICE BAKED POTATO CASSEROLE
Serves: 4

3 large potatoes
¼ cup milk
6 tablespoons butter
1 (8 oz.) carton sour cream
1 cup shredded sharp Cheddar
cheese, divided

½ cup chopped green onions
6 slices crisp bacon, crumbled
Salt and pepper to taste

Preheat oven to 400°.   2 quart casserole
Wrap potatoes in foil; bake for 1 hour, or until done. When done, remove from foil; place in a large mixing bowl. Chop into pieces; add milk, butter, and sour cream. Blend with a mixer. Add ½ cup cheese, green onions, bacon, and seasonings. Mix until blended. Place in casserole; top with remaining cheese. Bake for 30 minutes, or until hot.

This recipe is good with potatoes peeled or unpeeled after baking.

VALLEY SEASONED POTATOES
Serves: 8-10

½ cup margarine
1 heaping tablespoon seasoned
   salt mixture

10 medium red potatoes,
   unpeeled and thinly sliced

Preheat oven to 350°.   2 quart casserole dish
Melt margarine; add seasoned salt mixture. Dip potatoes in mixture; arrange in overlapping rows in glass dish; bake for 1 hour.

Seasoned Salt Mixture:
2 tablespoons celery salt
2 tablespoons garlic powder
2 tablespoons onion salt
1½ teaspoons paprika

1¼ teaspoons chili powder
½ teaspoon pepper
⅛ teaspoon cayenne pepper

Combine all spices; sift together 3 or 4 times. Store in jar with tight lid.

BROWN BUTTER RICE

Serves: 4

3 cups chicken broth
Juice of 1 lemon

1 cup white, long grain rice
¼ cup butter

Bring chicken broth and lemon juice to a boil in medium saucepan. Add rice, return to a boil; reduce heat. Cover and simmer until liquid is absorbed, about 20 minutes. In another pan, over medium heat, melt and simmer butter until it becomes lightly browned, but do not burn. Stir into cooked rice.

BROWN RICE PILAF

Serves: 4

4 tablespoons unsalted butter
or 2 tablespoons butter and 2
tablespoons oil
¼ cup chopped onion
1 cup brown rice
2½ cups chicken stock or water
1 teaspoon salt

¼ teaspoon pepper
½ cup golden raisins
½ cup dry white wine
¾ cup slivered almonds, toasted
½ cup chopped mint or
coriander

In a medium saucepan, sauté the onions in butter until translucent. Add rice; cook over low heat until grains are coated with butter and have browned a little (about 3 minutes). Add stock or water; stir, and season with salt and pepper. Bring to a boil. Cover and simmer 45 minutes, or until all liquid is absorbed. Meanwhile, plump raisins in white wine. Remove rice from heat, add almonds, raisins, and chopped mint or coriander to rice; mix well. Turn into a serving bowl; garnish with mint or coriander.

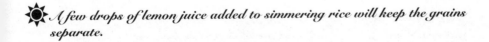

A few drops of lemon juice added to simmering rice will keep the grains separate.

CONFETTI RICE

Serves: 8-10

12 tablespoons of butter, divided
1 cup slivered almonds
1 cup shredded carrots,
  blanched for 1 minute

1 cup frozen green peas, rinsed
  and drained
4 cups cooked white rice
1 teaspoon salt

In large skillet, melt 2 tablespoons butter; stir fry almonds for 1 minute and remove. Melt 2 tablespoons butter; stir fry the carrots, and remove. Melt 2 tablespoons butter; stir fry the peas for 1 minute, and remove. Melt 6 tablespoons butter; add the rice, stir fry rice for 1 minute, then add the other ingredients. Mix well and serve.

WILD RICE

Serves: 6

1 cup wild rice, soaked
  overnight
1 (10½ oz.) can consommé
4 tablespoons butter
¾ pound mushrooms, sliced
1½ cups finely chopped celery
1 bunch green onions, sliced

1 (6 oz.) can water chestnuts,
  drained and chopped
3 tablespoons chopped parsley
¼ cup slivered almonds, toasted
½ cup vermouth
Butter

Preheat oven to 350°.   2 quart casserole

Butter casserole and set aside. Rinse wild rice well; drain. Combine wild rice with consommé in a large saucepan; simmer, covered, until liquid is absorbed (about 30 minutes). In a large skillet, melt butter; sauté vegetables until limp. Add parsley and almonds; mix thoroughly. Combine vegetable mixture with cooked rice; place in casserole. Refrigerate until ready to use. When ready to heat, add vermouth, dot with butter, and cover. Bake for 30-40 minutes.

Rubbing the inside of the cooking vessel with vegetable oil will also prevent noodles, spaghetti, and similar starches from boiling over.

SPINACH CASSEROLE

Serves: 8

1 onion, chopped
1 (10 oz.) box frozen chopped
   spinach, thawed and drained
½ cup butter
1 (10 oz.) can cream of chicken
   soup
1 (8 oz.) carton sour cream

2 (4 oz.) cans chopped green
   chilies
2 (8 oz.) mixed packages
   grated Cheddar and Monterey
   Jack cheese
12 flour tortillas, torn into pieces

Preheat oven to 350°.   2 quart casserole dish

Lightly grease casserole dish. In skillet, sauté onion and spinach in butter; mix in soup, sour cream, chilies, one package of cheese, and tortillas. Pour into dish and top with other package of cheese. Bake 15 minutes.

SPINACH CRUNCH

Serves: 10

4 tablespoons margarine
3 (10 oz.) packages frozen
   chopped spinach, thawed and
   drained

4 ounces cream cheese
½ teaspoon nutmeg
1 cup Parmesan cheese
1 cup finely chopped pecans

Preheat oven to 350°.   11 x 7 casserole dish

Spray dish with nonstick spray. Melt margarine in skillet and add cream cheese. Stir until cheese is softened; add spinach and nutmeg, mixing well. Pour mixture into dish and cover with Parmesan cheese. Sprinkle pecans on top of cheese and bake 40 minutes. Let stand 5 minutes before cutting into squares. Serve immediately.

SPINACH SOUFFLE

Serves: 6

1 package Stouffers spinach souffle, thawed
2 eggs
3 tablespoons milk
¼ cup minced onion
½ cup sliced mushrooms
¾ cup cooked Italian sausage
¾ cup grated Swiss cheese
1 unbaked 9-inch pie shell

Preheat oven to 400°.

Mix all ingredients in large mixing bowl. Pour into pie shell. Bake for 35 minutes.

SQUASH CASSEROLE

Serves: 6-8

6 yellow squash, diced
½ cup chopped onion
½ cup butter or margarine
Salt and pepper to taste
1 (16½ oz.) can whole kernel corn, drained
½ cup chopped green chilies, drained
1 cup grated cheddar cheese

Preheat oven to 350°.   13 x 9 baking pan

Prepare baking pan. In a large skillet, sauté squash and onion in butter until tender. Add salt, pepper, corn and green chilies. Place in baking pan and top with cheese. Bake until cheese is melted.

BAKED SQUASH CASSEROLE

Serves: 8-10

3 pounds fresh squash
½ cup margarine
1 (8 oz.) package Pepperidge
Farm Stuffing Mix
1 onion, chopped
1 cup chopped green bell
pepper

1 carrot, grated
1 (10 oz.) can cream of chicken
soup
1 cup grated Cheddar cheese
1 (8 oz.) carton sour cream

Preheat oven to 350°.   13 x 9 casserole dish

Cook squash in salted water. Drain well. Melt margarine and mix in ½ of stuffing mix. Combine squash, onion, bell pepper, carrot, soup, Cheddar cheese, and sour cream. Put in dish and top with remaining stuffing mix. Bake covered for 45 minutes.

BUTTERNUT SQUASH SOUFFLÉ

Serves: 8-10

2 cups cooked, mashed
butternut squash
3 eggs, beaten
⅓ cup butter, melted

½ cup milk
¾-1 cup sugar
½ teaspoon ground ginger
½ teaspoon vanilla extract

Preheat oven to 375°.   13 x 9 baking pan

Lightly grease baking pan. Bake squash for about 1 hour, or until tender, or boil squash until tender. Peel and remove seeds. Mash and mix with remaining ingredients. Reduce oven to 350°. Pour into pan and bake 1 hour.

 To dewax cucumbers, green peppers, and rutabagas, soak for 5 minutes in a quart of tap water containing 10-15 drops of liquid detergent. Then wash with 1 tablespoon of herbal or brown vinegar. Rinse in fresh tap water and rub with a towel. Do not soak in salted water.

GREEN AND GOLD QUICHE

Serves: 6-8

1 (10 inch) pastry shell, unbaked
2 medium zucchini, thinly sliced
2 medium yellow squash, thinly sliced
½ medium onion, chopped
2 green onions, sliced
1 clove garlic, minced
1 medium tomato, peeled and chopped

1 green bell pepper, finely chopped
¾ teaspoon salt
¼ teaspoon pepper
½ teaspoon basil
2 tablespoons butter or margarine, melted
3 eggs, beaten
½ cup whipping cream
¼ cup grated Parmesan cheese

Preheat oven to 450°.   10 inch pie pan

Prick bottom and sides of unbaked pastry shell. Bake for 8 minutes. Set aside to cool. Reduce oven to 350°. Combine vegetables, salt, pepper, basil, and butter in a large skillet. Sauté until vegetables are tender. Spoon vegetables evenly into pastry shell. Combine eggs and cream, mixing well; pour over vegetables. Sprinkle with cheese and bake for 30 minutes.

ITALIAN SQUASH

Serves: 5

4 medium zucchini squash
1 medium onion, diced
2 teaspoons olive or salad oil
1 medium tomato, diced
Salt and pepper to taste

⅛ teaspoon garlic powder
¼ cup chopped fresh parsley
¼ teaspoon oregano
¼ teaspoon basil
2 tablespoons water

Slice zucchini in food processor. Spray skillet with non-stick spray. In hot oil, sauté onion until golden brown. Add tomato and zucchini, along with rest of ingredients. Cover skillet. Cook over low heat about 20 minutes. More cooking will diminish crispness.

 Scalding tomatoes, peaches, or pears in boiling water before peeling makes them easier to peel and the fruit skins slip right off.

RIO GRANDE VALLEY SQUASH CASSEROLE Serves: 6

2 pounds diced yellow squash
2 sliced medium onions
1 teaspoon salt (optional)
½ teaspoon pepper
⅛ teaspoon garlic powder
½ teaspoon ground nutmeg

2 eggs, slightly beaten
5 ounces evaporated milk
1 tablespoon sugar
½ cup Italian bread crumbs
¼ cup butter

Preheat oven to 350°.   11 x 7 baking pan

Lightly butter baking pan. Cook squash and onions in small amount of salted water until tender. Drain and mash until lumpy. Add all remaining ingredients except butter and mix well. Place in pan. Top with more bread crumbs and dot with butter. Bake about 20 minutes.

BOURBON SWEET POTATOES   Serves: 10-12

6 cups sliced sweet potatoes,
   cooked or canned
4 tablespoons butter or
   margarine

Milk

Preheat oven to 350°.   3 quart casserole dish

Spray casserole with non-stick spray. Blend potatoes, butter or margarine, and enough milk until it reaches the consistency of thick mashed potatoes. Place potatoes in casserole, making a nest in center large enough for 1 cup of filling.

Filling:

4 tablespoons butter or
   margarine
1 cup light brown sugar, firmly
   packed

1 cup warm half-and-half
¼ cup bourbon
Marshmallows

Cream butter or margarine and sugar in a small saucepan. Gradually add half-and-half. Stir over low heat until mixture boils. Remove from heat; add bourbon. Pour filling in nest; bake 30 minutes. Remove from oven; cover top of potatoes with marshmallows. Bake 10 minutes more.

SWEET POTATO SOUFFLÉ
Serves: 6

3 cups mashed, cooked sweet
potatoes
½ cup milk
4 tablespoons butter
½ cup sugar
2 teaspoons vanilla extract
¼ teaspoon salt

2 eggs
1 cup brown sugar, firmly
packed
½ cup self-rising flour
½ cup chopped pecans
4 tablespoons butter

Preheat oven to 350°. 8 x 8 casserole dish

Grease casserole dish. Combine sweet potatoes, milk, butter, sugar, vanilla, salt, and eggs; place in casserole. Mix brown sugar, flour, pecans, and butter; sprinkle on top. Bake for approximately 30 minutes.

TROPICAL SWEET POTATO CASSEROLE
Serves: 6-8

1 (17 oz.) can sweet potatoes
2 medium size ripe bananas
½ lemon, juiced
¼ teaspoon salt

4 tablespoons butter
8 tablespoons honey
Mini-marshmallows

Preheat oven to 350°. 9 x 9 baking pan

Drain ⅔ liquid from canned sweet potatoes. Mash potatoes with remaining juice and bananas; sprinkle with lemon juice and salt. Add butter and honey; mix well. Pour mixture in pan; top with mini-marshmallows. Bake until marshmallows are browned and melted.

TOMATOES FLORENTINE

Serves: 4

4 medium tomatoes
¼ cup chopped onion
2 tablespoons margarine
1 (10 oz.) package frozen
  spinach, thawed

¼ cup chicken broth
2 tablespoons flour
  Cayenne pepper, to taste
½ cup buttered bread crumbs

Preheat oven to 375°.   9 x 9 baking dish

Cut tops from tomatoes and remove centers. Salt shells then turn upside down on paper towels to drain. Cook onion in margarine; stir in flour and chicken broth. (May not need all chicken broth.) Add spinach and cook for 3-5 minutes. Remove from heat and add cayenne pepper; fill the prepared tomato shells with spinach mixture. Sprinkle bread crumbs on top. Place in a baking dish and bake for 20-25 minutes.

ITALIAN MARINATED VEGETABLES

Serves: 12

1 (8 oz.) can button
  mushrooms, drained
1 green bell pepper, cut into
  ½ inch strips
1 red or yellow bell pepper, cut
  into ½ inch strips
1 carrot, cut lengthwise into
  eighths
2 cups uncooked cauliflower
  flowerets
1 (1 lb.) can artichoke hearts,
  drained and cut in half

6 green onions, chopped
½ cup stuffed olives
1 jar salad green peppers
½ cup sliced radishes
1½ cups wine vinegar
4½ cups olive oil or vegetable oil
1 teaspoon sugar
½ teaspoon salt
1 teaspoon pepper
2 teaspoons crushed oregano

Combine all vegetables in large bowl. Heat vinegar; stir in seasonings. Cook slightly, combine with oil, pour over vegetables; mix well. Cover and refrigerate for 24 hours before serving, stirring occasionally.

This keeps well and can be made several days ahead.

VEGETABLE CASSEROLE

Serves: 8-10

1 (10 oz.) package frozen
  chopped broccoli, cooked
1 (10 oz.) package frozen green
  beans or lima beans, cooked
2 cans sliced water chestnuts
1 (10¾ oz.) can cream of
  mushroom soup

1 package dried onion soup mix
1 cup sour cream
3 cups Rice Krispies
6 tablespoons butter

Preheat oven to 350°.   13 x 9 baking pan

Drain cooked vegetables. Combine vegetables, water chestnuts, soup, soup mix, and sour cream. Pour into casserole dish. Sauté Rice Krispies in butter. Top vegetables with Rice Krispies and bake, uncovered, 25 minutes.

VEGETABLE MEDLEY

Serves: 12

2 heads broccoli flowerets
2 heads cauliflower flowerets
1 (1 lb.) bag carrots, sliced
1 green bell pepper, sliced
3 stalks celery, sliced
1 (10¾ oz.) can cream of
  potato soup
1 (10¾ oz.) can cream of celery
  soup

1 (12 oz.) can evaporated milk
½ cup butter or margarine,
  melted
1 (4 oz.) can mushrooms
  Shredded Cheddar cheese
  or French-fried onion rings

Preheat oven to 350°.   3 quart casserole dish

Grease casserole dish. Steam the vegetables. While vegetables are cooking, mix the cans of soup, milk, butter or margarine, and mushrooms. Place cooked vegetables in casserole and top with soup mixture. Sprinkle with cheese or onion rings and bake, uncovered, 30 minutes or until heated through.

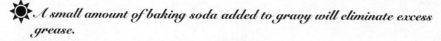 *A small amount of baking soda added to gravy will eliminate excess grease.*

Sweets

Refer to Mexican Cuisine section for additional recipes in this category.

Adult Literacy
Working on a one-on-one basis, our volunteers, in cooperation with the McAllen Public Library, teach the basics of reading, writing, and speaking English to individuals wishing to learn or improve their English skills. This program solicits prospective tutors and promotes an awareness of the literacy situation throughout the Rio Grande Valley.

 These recipes are pictured on the previous page.

220

CHOCOLATE CHERRY SQUARES

Serves: 24

Cake:

1 package devil's food cake mix with pudding	2 eggs
	1 teaspoon almond extract
½ cup sour cream	1 (21 oz.) can cherry pie filling

Preheat oven to 350°. 13 x 9 baking pan

Grease and flour pan. Mix ingredients together with a wooden spoon. Pour into prepared pan. Bake for 45-50 minutes or until toothpick comes out clean. Watch closely the last 20 minutes so edges do not burn. When done, set cake aside in the pan; prepare frosting for the warm cake.

Frosting:

1 (6 oz.) package semi-sweet chocolate chips	½ teaspoon almond extract
	3-5 tablespoons milk, or more if needed to spread
4 tablespoons margarine	
2½ cups powdered sugar	

Melt chocolate chips and margarine together. Stir in sugar, almond extract, and enough milk to achieve spreading consistency. Spread warm frosting over warm cake in the pan.

PIÑA COLADA CAKE

Serves: 10-12

1 box yellow cake mix	2 tablespoons rum
1 (14 oz.) can sweetened condensed milk	1 large can or bag (7 oz.) shredded coconut
4 ounces of a 15 ounce can of Coco Lopez (real cream of coconut)	1 large container Cool Whip

Preheat oven to 350°. 13 x 9 baking pan

Grease and flour pan. Bake cake according to package directions. Remove from oven and prick top of cake with toothpick. Spoon milk over warm cake. Mix cream of coconut and rum; pour over cake. Let cool and refrigerate covered. Before serving, spread Cool Whip over top of cake and sprinkle with coconut.

THIRTY MINUTE CHOCOLATE CAKE

Serves: 20

2 cups flour	½ cup buttermilk
2 cups sugar	½ cup margarine, melted
2 eggs	½ cup oil
1 teaspoon vanilla extract	3 tablespoons cocoa
1 teaspoon baking soda	1 cup boiling water

Preheat oven to 350°.   13 x 9 baking pan

Grease and flour pan. In a large mixing bowl, combine all ingredients. Mix thoroughly. Pour into pan and bake for 25-30 minutes. Ice cake while warm.

Icing:

½ cup margarine	1 (16 oz.) box powdered sugar
3 tablespoons cocoa	1 teaspoon vanilla extract
6 tablespoons half-and-half	1 cup chopped pecans

In a small saucepan, melt margarine, cocoa, and half-and-half. Add powdered sugar and mix until smooth. Add vanilla and pecans and mix well. Ice cake in pan directly from oven.

LEMON CAKE

Serves: 12

1 (3 oz.) package lemon gelatin	4 eggs
¾ cup boiling water	¾ cup oil
1 box lemon supreme cake mix	1 tablespoon lemon extract

Preheat oven to 350°.   Tube pan

Grease pan. Dissolve gelatin in boiling water. Cool. Combine the cake mix, eggs (beating well after each egg), oil, and lemon extract. Add the lemon gelatin and beat well with mixer. Pour batter into pan and bake for 40-50 minutes. Remove cake from oven and take out of pan. Prick top with toothpick and spoon glaze over top of cake.

Glaze:

¼ cup fresh lemon juice	1½ cups powdered sugar

Mix ingredients together and spoon over top of warm cake.

 Freeze cake before icing to prevent crumbs mixing with icing.

COCONUT AMARETTO CAKE

Serves: 12

Cake:

- 5 eggs
- 2 cups granulated sugar
- 1 cup butter or margarine, softened
- 2 cups flour
- 1 teaspoon baking soda
- 1 cup buttermilk
- 6 ounces angel flake coconut
- ½ cup amaretto
- 4 heaping tablespoons mayonnaise

Preheat oven to 350°.   3 (8 inch) cake pans

Grease and flour cake pans. Separate eggs; beat egg whites until stiff peaks form. Set aside. Cream together sugar, softened butter, and egg yolks. Add flour, baking soda, buttermilk, coconut, amaretto, and mayonnaise. Gently fold egg whites into mixture. Divide batter evenly among pans. Bake for 25-30 minutes, or until cake tests done. Cool before removing from pans.

Frosting:

- 1 (8 oz.) package cream cheese
- ½ cup butter
- 1 (1 lb.) box of powdered sugar
- 2 teaspoons amaretto
- 8 ounces angel flake coconut
- 1 cup pecans, chopped

Cream first four ingredients together. Frost between cake layers; sprinkle with coconut and pecans. (Nuts and coconut may be mixed into frosting.)

HURRICANE CAKE

Serves: 12-16

- 1 cup chopped pecans
- 1⅓ cups flaked coconut
- 1 German chocolate cake mix
- 1 (8 oz.) package cream cheese, softened
- 1 (16 oz.) box powdered sugar
- 1 teaspoon vanilla extract

Preheat oven to 350°.   13 x 9 baking pan

Grease and flour pan. Mix pecans and coconut together; distribute on the bottom of the prepared pan. Following directions on box of cake mix, mix and pour on top of pecans and coconut. In another bowl, using a mixer, combine cream cheese, powdered sugar, and vanilla. Distribute this over the cake mixture. Bake approximately 50 minutes.

MANDARIN ORANGE CAKE

Serves: 12-16

1 box yellow cake mix, without pudding
4 eggs
1 cup oil

1 (15 oz.) can mandarin oranges, drained
½ teaspoon orange extract

Preheat oven to 350°. 3 round (8 inch) cake pans
Grease and flour cake pans. Mix ingredients well, adding eggs one at a time. Bake in 3 prepared pans for 15-18 minutes. Cool. (If using a different size pan, check directions on back of cake mix for cooking time.) Frost with Chocolate Buttercream Frosting or Pineapple Coconut Frosting.

Chocolate Buttercream Frosting:

6 tablespoons butter or margarine, softened
¾ cup cocoa
2⅔ cups unsifted powdered sugar

⅓ cup milk plus 1 teaspoon, if needed
1 teaspoon vanilla extract

Cream butter in small mixing bowl. Add cocoa and powdered sugar alternately with milk. Beat to spreading consistency. Add additional teaspoon of milk, if necessary. Blend in vanilla. Frost cake.

Pineapple Coconut Frosting:

1 (20 oz.) can crushed pineapple
1 (5.9 oz.) package instant vanilla pudding

1 (3½ oz.) can coconut
1 (12 oz.) container refrigerated dairy topping.

Mix first 3 ingredients well. Fold into topping. Frost cake as desired.

 Sprinkle cake plate with granulated sugar to prevent cake from sticking when serving.

ORANGE CHIFFON CAKE

Serves: 12-16

2 cups sifted flour
1½ cups sugar
3 teaspoons baking powder
1 teaspoon salt
½ cup oil
5 egg yolks, unbeaten

Juice of 2 oranges and water to make ¾ cup liquid
2 tablespoons grated orange rind
1 cup (7-8) egg whites
½ teaspoon cream of tartar

Preheat oven to 325°.   Tube pan

Grease pan. In a small bowl, mix first 8 ingredients. In a large bowl, mix last two ingredients and whip until whites form very stiff peaks. Gently fold ingredients of small bowl into large bowl. Pour into pan; bake for 65-70 minutes. Remove from oven; invert on glass soda bottle to cool. Cool completely before icing.

Icing:

½ cup butter or margarine
3 tablespoons flour
¼ teaspoon salt
½ cup fresh orange juice

3 cups sifted powdered sugar
Cold water
1 tablespoon grated orange rind

In a medium saucepan, melt butter. Remove from heat. Blend in flour and salt. Slowly stir in orange juice. Bring to a boil, stirring constantly. Boil for 1 minute and remove from heat. Stir in powdered sugar. Set saucepan in cold water. Beat until smooth and spreadable. Stir in grated orange rind.

 Cream whipped ahead of time will not separate if you will add 1/4 teaspoon unflavored gelatin per cup of cream.

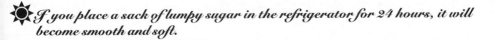

If you place a sack of lumpy sugar in the refrigerator for 24 hours, it will become smooth and soft.

SPECIAL SPICE CAKE

Serves: 12

3 cups sifted flour
1½ cups sugar
1½ teaspoons baking powder
1½ teaspoons cinnamon
¾ teaspoon nutmeg
¾ teaspoon cloves
¾ teaspoon allspice
¾ teaspoon salt

1 (8 oz.) can tomato sauce
1½ teaspoons baking soda
2 eggs, slightly beaten
¾ cup oil
1 cup chopped nuts
1½ cups golden raisins, optional
½ cup orange or pineapple juice
Powdered sugar

Preheat oven to 350°.   10 inch Bundt or tube pan

Lightly grease pan. In a large mixing bowl, combine flour, sugar, baking powder, spices, and salt. Thoroughly mix tomato sauce and soda in a small bowl: add to flour mixture. Stir in eggs, oil, nuts, raisins, and fruit juice. Mix well. Pour into prepared pan. Bake for 45-55 minutes. Cool in pan 15 minutes before turning out on serving plate. Dust top with powdered sugar.

POUND CAKE WITH LEMON SAUCE

Serves: 12

1 cup butter
1¾ cups sugar
5 large eggs

2 cups sifted cake flour
2 teaspoons vanilla extract
1 teaspoon almond extract

Preheat oven to 350°.   Bundt pan

Grease and flour pan. Cream butter and sugar very well. Add eggs one at a time. Add cake flour, vanilla, and almond extract. Pour mixture into pan. Bake for 25 minutes or until done.

Lemon Sauce:

½ cup sugar
4 teaspoons cornstarch
1 cup water
1 tablespoon butter

⅛ teaspoon salt
2 egg yolks, well beaten
5-6 tablespoons lemon juice
1 tablespoon grated lemon rind

In a heavy saucepan, mix cornstarch and sugar well. Stir in water, butter, and salt. Cook over medium heat until it just begins to thicken. Whisk in egg yolks slowly. Add lemon juice and rind. Cook slowly and stir constantly until medium thick, sauce will thicken as it cools. Pour over cake. Extra sauce may be stored in refrigerator.

WHIPPED CREAM POUND CAKE

Serves: 12

1 cup butter, softened
3 cups sugar
6 eggs
½ pint whipping cream

¼ teaspoon cream of tartar
1½ teaspoons vanilla extract
3 cups cake flour

Preheat oven to 350°.   Bundt pan

Grease pan. Cream butter and sugar in a small bowl. Add eggs, beating well after each addition. Mix whipping cream, cream of tartar, and vanilla in a small bowl. Add ingredients of small bowl to large bowl, alternating with the cake flour. Beat well. Pour batter into prepared pan. Bake for 60-70 minutes.

Glaze:

1½ cups powdered sugar
2 tablespoons water

½ teaspoon vanilla extract
Heath Brickle Bits

Mix all ingredients and pour over warm cake.

BLUEBERRY POUND CAKE

Serves: 12-14

1 cup butter
2 cups sugar
4 eggs
1 teaspoon vanilla extract
2¾ cups flour

1 teaspoon baking powder
½ teaspoon salt
2 cups blueberries
¼ cup flour

Preheat oven to 350°.   Bundt pan

Spray the pan with a vegetable cooking spray, then coat with sugar. In a large bowl, cream the butter and sugar. Add the eggs, one at a time, beating well after each addition; add the vanilla, 2¾ cups flour, baking powder, and salt. In another bowl, toss the berries and the ¼ cup flour, then fold berries into the batter. Pour batter into prepared pan and bake for 1¼ hours. Remove from oven, cool 15 minutes, then remove from pan. This recipe is better prepared a day ahead.

DELICIOUS CHEESECAKE

Serves: 12

Crust:

2 cups crushed graham crackers

1 cup butter, melted

Preheat oven to 400°.

10 inch springform pan

Set aside 2-3 tablespoons of crumbs. Mix remaining crumbs with butter. Press into pan.

Filling:

2 pounds cream cheese, softened

½ cup sugar

2 eggs, lightly beaten

2 tablespoons cornstarch

1 cup sour cream

2 tablespoons vanilla extract

Blend cream cheese with sugar; add eggs, mix well. Add cornstarch; mix well. Add sour cream and vanilla; mix well. Pour over crust. Bake for 10 minutes; reduce heat to 225°; bake for 40 minutes longer. Turn oven off, prop door open; let cool in oven 2 to 3 hours. Sprinkle top with reserved crumbs or use your favorite sauce as a topping.

FUDGE TRUFFLE CHEESECAKE

Serves: 6-8

Crust:

1½ cups vanilla wafer crumbs

½ cup powdered sugar

⅓ cup cocoa

⅓ cup butter, melted

Preheat oven to 300°.

9 inch springform pan

Combine crumbs, sugar, cocoa, and butter. Press into bottom of pan. Set aside.

Filling:

2 cups semi-sweet chocolate chips

3 (8 oz.) packages cream cheese, softened

1 (14 oz.) can sweetened condensed milk

4 eggs

2 teaspoons vanilla extract

Over low heat melt chocolate chips, stirring constantly. In a large mixing bowl, beat cream cheese until fluffy. Gradually beat in sweetened condensed milk until smooth. Add melted chocolate, eggs, and vanilla; mix well. Pour into crust. Bake 65 minutes or until center is set. Cool, then keep refrigerated.

May be garnished with shaved chocolate and whipped cream.

PUMPKIN CHEESECAKE

Serves: 12

Crust:

2 cups crushed graham
crackers

¼ cup sugar
6 tablespoons butter, melted

Preheat oven to 350°.   9 inch springform pan

Using a food processor, crumble the graham crackers. Add melted butter and sugar. Press into pan.

Filling:

3 (8 oz.) packages cream
cheese, room temperature
¾ cup sugar
¼ cup packed brown sugar
5 eggs
¼ cup heavy cream

1 (16 oz.) can pumpkin
1 teaspoon cinnamon
½ teaspoon nutmeg
¼ teaspoon ground cloves
¼ teaspoon ginger

Beat cream cheese, sugar, and brown sugar; mix until light and fluffy. Add eggs, one at a time, beating well. Add remaining ingredients on medium speed mixing thoroughly. Pour batter into prepared crust; bake for 1 hour.

Topping:

6 teaspoons butter, melted
½ cup brown sugar (1 cup for
sweeter taste)

1 cup chopped walnuts

Mix together melted butter, brown sugar, and walnuts. Sprinkle topping on cheese cake; bake 15 minutes longer. Let cool in a draft-free area. Run a small knife around the edge of springform pan to prevent the cheesecake from cracking in the middle. When completely cool, cover and refrigerate.

 *Use dental floss to cut cheesecake neatly; then pull floss out sideways.*

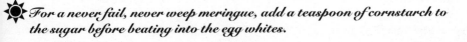 *For a never fail, never weep meringue, add a teaspoon of cornstarch to the sugar before beating into the egg whites.*

ORANGE BLOSSOM CHEESECAKE

Serves: 10-12

Crust:

3 cups ground gingersnap cookies

6 tablespoons butter or margarine, melted

2 teaspoons grated orange peel

9 inch springform pan

Mix above ingredients in small bowl until well blended. Press crumbs into bottom and sides of pan. (Crust will not go to top of cheesecake.)

Filling:

1½ cups fresh orange juice

⅓ cup thinly sliced fresh ginger, unpeeled

4 (8 oz.) packages cream cheese, softened

⅔ cup sugar

2 tablespoons grated orange peel

1 tablespoon vanilla extract

6 ounces imported white chocolate, melted

4 large eggs

Preheat oven to 350°.

Boil orange juice and ginger in heavy saucepan until reduced to 3 tablespoons. Using electric mixer, beat cream cheese, sugar, orange peel, and vanilla until smooth. Strain juice and add to cream cheese mixture. With mixer running, add chocolate and beat until well blended. Reduce speed to low and add eggs one at a time, beating until just combined. Pour batter over crust. Bake until top is dry, approximately 50 minutes (cheesecake will jiggle when shaken). Transfer to rack and cool. Cover and chill overnight.

Topping:

4 cups water

2 cups sugar

2 seedless oranges, unpeeled and cut into very thin slices

Combine water and sugar in large skillet. Stir over medium heat until sugar dissolves. Simmer 5 minutes. Add orange slices one at a time and adjust heat so syrup bubbles only around edges of pan. Cook 1 hour. Turn over top layer of oranges and cook until all are translucent and orange peels are tender, about 1 hour longer.

Arrange oranges in a single layer on a rack that has been covered with wax paper. Let dry 1 hour.

Remove sides from springform pan and overlap orange slices atop cheesecake.

OREO CHEESECAKE

Serves: 12

Crust:

30 Oreo cookies

5 tablespoons unsalted butter, melted

Preheat oven to 425°.

9 inch or 10 inch springform pan

Lightly butter pan. Break up cookies and place in a food processor fitted with a metal blade; process until crumbs. Add butter and mix until blended. Pour into pan. Press evenly over bottom and two-thirds up sides of pan. Refrigerate while preparing filling.

Filling:

4 (8 oz.) packages cream cheese, at room temperature

1¼ cups sugar

2 tablespoons cake flour

4 large eggs, at room temperature

3 large egg yolks, at room temperature

⅓ cup whipping cream

1 teaspoon vanilla extract

18 Oreo cookies, chopped into quarters

In a large bowl, beat cream cheese with mixer on medium speed until smooth. Scrape down sides of bowl and add sugar, beating until mixture is fluffy—about 3 minutes. Mix in flour. Continue beating and add eggs and yolks; mix until smooth. Beat in whipping cream and vanilla. Pour half of the batter into prepared crust. Sprinkle with quartered Oreos. Pour remaining batter over cookies. Place pan on baking sheet and bake for 15 minutes. Reduce oven to 225° and bake for an additional 50 minutes, or until set. Remove cake from oven and increase oven temperature to 350°.

Topping:

2 cups sour cream

¼ cup sugar

1 teaspoon vanilla extract

In a small bowl, stir together sour cream, sugar and vanilla. Spread sour cream mixture evenly on top of cake. Return to 350° oven and bake for 7 minutes or until sour cream begins to set. Remove from oven and cool in draft-free place to room temperature. Cover cake in pan and refrigerate several hours or overnight.

This cake may be refrigerated up to 3 days or it may be frozen for 1 month as long as it is covered well. Thaw in refrigerator overnight before serving. Before serving, remove sides of springform pan. Serve chilled.

TRIPLE LAYER CHEESECAKE

Serves: 10-12

Chocolate crust:

1 (8½ oz.) package chocolate wafer cookies, crushed (about 2 cups)

¼ cup sugar

5 tablespoons butter or margarine, melted

Preheat oven to 325°.   9 inch springform pan

Combine cookie crumbs, sugar, and butter in a medium bowl; blend well. Press into the bottom and 2 inches up sides of pan. Set aside.

Use 2 cups crushed Oreo cookies, minus filling if you cannot find chocolate wafers.

Cheesecake:

2 (8 oz.) packages cream cheese, softened and divided

½ cup sugar, divided

3 eggs, divided

1 teaspoon vanilla extract, divided

2 (1 oz.) squares semi-sweet chocolate, melted

1⅓ cups sour cream, divided

⅓ cup firmly packed dark brown sugar

1 tablespoon all-purpose flour

½ cup finely chopped pecans

6 ounces cream cheese, softened

¼ teaspoon almond extract

Chocolate Glaze (recipe follows)

Chocolate leaves (optional), see page 247

Combine 1 (8 oz.) package cream cheese and ¼ cup sugar; beat until fluffy. Add 1 egg and ¼ teaspoon vanilla; blend well. Stir in melted chocolate and ⅓ cup sour cream. Spoon over chocolate crust. Combine remaining (8 oz.) package cream cheese, brown sugar, and flour; beat until mixture is fluffy. Add 1 egg and ½ teaspoon vanilla; blend well. Stir in pecans. Spoon gently over chocolate layer. Combine 6 ounces cream cheese and remaining ¼ cup sugar; beat until fluffy. Add egg; blend well. Stir in remaining 1 cup sour cream, ¼ teaspoon vanilla, and almond extract. Spoon gently over praline (previous) layer. Bake for 1 hour; turn oven off and leave cheesecake in oven for 30 minutes; open door of oven and leave cheesecake in oven an additional 30 minutes. Cool. Chill 8 hours; remove from pan. Spread warm chocolate glaze over cheesecake. Garnish with chocolate leaves if desired.

(Continued on next page)

(Triple Layer Cheesecake, continued from previous page)

Chocolate Glaze:

6 (1 oz.) squares semi-sweet chocolate

4 tablespoons butter or margarine

¾ cup sifted powdered sugar

2 tablespoons water

1 teaspoon vanilla extract

Combine chocolate and butter in top of a double boiler; cook until melted. Remove from heat; stir in remaining ingredients until smooth. Spread over cheesecake while glaze is warm.

BUTTERMILK PRALINES

Yield: 2 dozen pieces

2 cups sugar

1 teaspoon baking soda

Pinch of salt

1 cup buttermilk

2 tablespoons butter

1½-2 cups pecans

1 teaspoon vanilla extract

Combine sugar, soda, salt, and buttermilk in a heavy 3 quart saucepan. Quickly bring to a boil, stirring constantly until mixture looks creamy (210°). Add butter and nuts, cooking over medium heat, stirring frequently to 230° (soft ball stage). Remove from heat, add vanilla. Beat until mixture loses its gloss, then drop quickly in mounds on foil. Work quickly; it thickens fast.

These are a very dark color.

A dip of the spoon or cup into hot water before measuring shortening or butter will cause the fat to slip out easily without sticking to the spoon.

CAPPUCCINO CARAMELS

Yield: 100 pieces

1 cup margarine or butter
2¼ cups brown sugar, firmly packed
1 (14 oz.) can sweetened condensed milk
1 cup light corn syrup
3 tablespoons instant coffee crystals

½-1 teaspoon finely shredded orange peel
1 cup chopped walnuts or pecans
1 teaspoon vanilla extract

8 x 8 baking pan

Butter and line pan with foil. In a heavy 3 quart saucepan melt margarine over low heat. Stir in brown sugar, sweetened condensed milk, corn syrup, coffee crystals, and orange peel. Carefully clip candy thermometer to the side of the saucepan. Cook over medium heat, stirring frequently, until thermometer registers 248° (firm-ball stage). Mixture should boil at a moderate, steady rate over the entire surface. Reaching firm-ball stage should take 15 to 20 minutes. Remove saucepan from heat; remove candy thermometer from saucepan. Immediately stir in nuts and vanilla. Quickly pour the caramel mixture into pan. When caramel is firm, use foil to lift it out of pan. Use a buttered knife to cut candy into ¾-1-inch squares. Wrap each piece in clear plastic wrap.

CANDIED GRAPEFRUIT PEEL

3 large grapefruit
1 (3 oz.) package red gelatin
2 cups water

1 cup sugar
½ cup red hot candies
Sugar

Cut grapefruit in half; remove pulp. Scrape as much of the white pectin away from the peel as possible. Cover with boiling water and boil 15 minutes. Remove peel; scrape remaining pulp and white pectin from peel. Cut in ¼ inch slices, return to fresh boiling water; boil an additional 15 minutes. Drain. Mix gelatin, water, red hots, and sugar until dissolved. Add grapefruit peel; cook over medium heat until almost all liquid is gone, about 50 minutes. Lift out; roll in sugar. Let dry on wax paper overnight.

 Store lemon, orange, and grapefruit rinds in the freezer; grate as needed.

PEANUT BUTTER BON BONS

Yield: 85 pieces

1 (18 oz.) jar creamy peanut
  butter
½ cup butter
2 cups powdered sugar, sifted
3 cups Rice Krispies

1 (8 oz.) Hershey chocolate bar
1 (6 oz.) package semi-sweet
  chocolate chips
⅛ pound paraffin

Cream peanut butter and butter well; add powdered sugar. Knead Rice Krispies into sugar mixture by hand. Roll into small balls in palms of hands. Set aside. In a double boiler, melt chocolates and paraffin together, stirring. Drop balls into chocolate to coat; remove with forks. Let dry on waxed paper.

Pickle forks work great for dipping.

CREAMY CARAMEL FUDGE

Yield: 24 to 48 squares

5 cups sugar, divided
2 cups milk
¼ cup butter

½ cup corn syrup
1 cup pecans

Jelly roll pan

Butter pan. Caramelize 1 cup of sugar by heating in a heavy nonferrous pan over low heat, stirring constantly with long handled spoon for 8 to 10 minutes, or until sugar has melted and is straw colored. When it is light brown, add milk; simmer until sugar dissolves. Add 4 cups sugar, butter, and corn syrup. Cook to 238°, stirring frequently. Cool to 180°, add pecans, and beat until thick. Pour into pan.

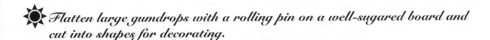

Flatten large gumdrops with a rolling pin on a well-sugared board and cut into shapes for decorating.

EASY VANILLA FUDGE

2 cups sugar
½ cup milk
1½ tablespoons corn syrup

¼ cup margarine (not butter)
1 teaspoon vanilla extract
1 cup chopped pecans

Jelly roll pan

Butter pan. In a heavy 2 quart saucepan, mix sugar, milk, corn syrup, margarine, and vanilla. Cook to 238°, stirring frequently. Cool to 170°; add pecans, and beat until thick. Pour into pan.

Add ¼ cup cocoa when cooking sugar mixture and proceed with recipe to make an easy chocolate fudge.

BROWNIE COOKIES

Yield: 3-4 dozen

1 (12 oz.) bag semi-sweet
chocolate morsels
1 stick butter
1 (14 oz.) can sweetened
condensed milk

1 cup flour
1 cup chopped pecans
Powdered sugar

Preheat oven to 350°.

Cookie sheet

Prepare cookie sheet. Melt chocolate chips, butter, and condensed milk in top of double boiler or in microwave. Add flour and pecans; mix well. Set aside for 15 minutes. Drop by teaspoonful on cookie sheet. Bake for 12 minutes. Sift powdered sugar over top of cookies while still hot.

CHOCOLATE-FILLED SNOWBALLS

Serves: 36-42

½ pound unsalted butter
¾ cup sugar
1 cup chopped pecans
1 tablespoon brandy

2 cups flour
1 (12 oz.) bag chocolate kisses
Powdered sugar

Preheat oven to 350°.

Cookie sheet

Cream butter, sugar, and pecans. Add brandy and flour. Roll dough into a ball; wrap in foil or plastic wrap. Refrigerate at least 30 minutes. Remove foil from kisses. Wrap each one inside a ball of dough, about 1 inch in diameter. Make sure each kiss is completely covered. Place on ungreased cookie sheet; bake about 12 minutes. Sprinkle with powdered sugar.

BRICKLE BARS

Yield: 16 bars

½ cup margarine or butter
2 (1 oz.) unsweetened
chocolate squares
1 cup sugar
2 eggs

1½ teaspoons vanilla extract
¾ cup flour
¾ cup almond brickle pieces
¾ cup miniature semi-sweet
chocolate pieces

Preheat oven to 350°.   8 x 8 baking pan

Lightly grease pan. In a saucepan over low heat, melt margarine and unsweetened chocolate squares. Remove from heat; stir in sugar. Add eggs and vanilla. Beat lightly with a wooden spoon until combined, being careful not to overbeat. Stir in flour. Spread batter in pan. Sprinkle almond brickle pieces and chocolate pieces evenly over batter. Bake for 30 minutes. Let cool; cut into bars.

Can be doubled.

BROWNIES

Yield: 2 dozen

Batter:

½ cup butter
3 squares unsweetened
chocolate
1½ cups sugar

3 eggs
1 cup flour
1 cup pecans
½ teaspoon salt

Preheat oven to 325°.   13 x 9 baking pan

Lightly grease and flour pan. Melt butter and chocolate squares together. Mix sugar and eggs together; add chocolate mixture. Combine flour, pecans, and salt; add to mixture. Spread batter in prepared pan. Bake for 25-30 minutes.

Icing:

3 tablespoons butter
2 tablespoons cocoa
1½ cups powdered sugar

2 tablespoons milk
1 teaspoon vanilla extract

Melt butter; add cocoa. Add sugar, milk, and vanilla. Stir until smooth. Frost brownies while warm. Cut into squares. Easy!

 Every time the door of the oven is opened, the temperature drops 25° - 30°.

CARAMEL BARS

Yield: 24 bars

1 (14 oz.) bag caramels
⅔ cup evaporated milk, divided
1 box German chocolate cake mix

½ cup margarine, melted
1 cup chocolate chips

Preheat oven to 350°.   13 x 9 baking pan

Lightly grease pan. Melt caramels in double boiler with ⅓ cup evaporated milk, stirring occasionally. In a medium mixing bowl, blend together dry cake mix, margarine, and ⅓ cup evaporated milk. Pour ½ of cake mixture into prepared pan. Bake for 6 minutes. Pour caramel mixture over cake; sprinkle with chips. Pour remaining cake mixture in dabs over filling. Bake for 15 minutes more. Cool completely before cutting.

CRUMBLE BARS

Serves: 6-8

Bars:
1½ cups flour
½ cup brown sugar, firmly packed
1 teaspoon baking powder

¼ teaspoon salt
5 tablespoons shortening
2 egg yolks, lightly beaten
½ cup pecans

Preheat oven to 350°.   11 X 7 baking pan

Lightly grease baking pan. Place flour, brown sugar, baking powder, and salt in food processor or electric mixer. Process until well blended. Add shortening and eggs; blend again. Pat mixture into pan. Pour topping over; sprinkle with nuts. Bake 20 minutes.

Topping:
2 egg whites
1 cup brown sugar, firmly packed

1 teaspoon vanilla extract

Beat egg whites until stiff, but not dry. Add brown sugar and vanilla. Mix well.

A slice of soft bread placed in the package of hardened brown sugar will soften it again in a couple of hours.

OATMEAL CRUNCHIES

Yield: 8 dozen

2 cups light brown sugar, firmly packed
2 cups granulated sugar
2 teaspoons vanilla extract
1½ cups oil
½ cup margarine or butter
4 eggs
4 cups flour

2 teaspoons baking soda
1 teaspoon salt
4 cups corn flakes, coarsely crumbled
1½ cups oatmeal
1-2 cups chopped pecans (optional)

Preheat oven to 350°.   Cookie sheet

Grease cookie sheet. In a large bowl, cream together sugars, vanilla, oil, and butter. Add eggs and continue to cream mixture. Add remaining ingredients, folding in corn flakes, oatmeal, and pecans last. Drop by teaspoonfuls onto cookie sheet; bake for 7-8 minutes.

SUGAR COOKIES

Yield: 3 dozen

1 cup unsalted butter
2 cups sugar
2 large eggs
4-5 cups flour

4 teaspoons baking powder
1 teaspoon salt
1 teaspoon vanilla extract

Preheat oven to 375°.   Cookie sheet

Grease cookie sheet. In a large bowl, cream butter, sugar, and eggs. In a medium bowl, sift together flour, baking powder, salt, and vanilla. Mix well, then divide in half and wrap in plastic wrap. (Although the mixture appears dry, you can shape it by hand into small balls.) Refrigerate for 1 hour. Roll dough out thick and cut desired shapes. Bake on cookie sheet for approximately 10 minutes. Best when cooled on a wooden cutting board and iced while hot. Use a powdered sugar icing.

PUMPKIN BARS

Serves: 24

Crust:
- 1 cup flour
- 1 cup quick cooking oats
- ½ cup light brown sugar, firmly packed
- ½ cup chopped pecans
- ½ cup butter or margarine

Preheat oven to 350°.   13 x 9 baking pan

Mix together flour, oats, sugar, pecans, and butter until crumbly. Press into bottom of baking pan. Bake for 15 minutes.

Filling:
- 1 (16 oz.) can pumpkin
- 2 eggs
- ¾ cup sugar
- ½ teaspoon salt
- 2 teaspoons pumpkin spice
- 1 (14 oz.) can evaporated milk

Beat together pumpkin, eggs, sugar, salt, spice, and milk; pour on top of baked crust. Return to oven; bake for 20 minutes.

Topping:
- ½ cup chopped pecans
- ½ cup light brown sugar, firmly packed
- 2 tablespoons butter or margarine

Mix together pecans, sugar, and butter or margarine; sprinkle over top. Bake an additional 15 to 20 minutes, or until lightly browned. Pumpkin bars are better if refrigerated.

SCOTCHEROOS

Yield: 5 dozen

- 1 cup sugar
- 1 cup light corn syrup
- 1 cup peanut butter
- 6 cups Rice Krispies cereal
- 1 (6 oz.) package chocolate chips
- 1 (6 oz.) package butterscotch chips

13 x 9 baking pan

Grease baking pan. In large saucepan, heat sugar and syrup over medium heat until it begins to boil. Remove from heat; stir in peanut butter. Add cereal; mix well. Pour into pan; press out evenly. Melt chips in double boiler. Spread over cereal. When set, cut into squares.

SOUR LEMON BARS

Yield: 40 squares

Crust:

1½ cups cake flour
¼ cup powdered sugar
⅛ teaspoon salt

½ cup chilled unsalted butter,
cut into pieces
½ teaspoon vanilla extract

Preheat oven to 350°.   9 x 9 baking pan

Line pan with heavy foil, extending 1 inch above 2 sides of pan. Butter 2 uncovered sides of pan. Position rack in center of oven. Combine flour, powdered sugar, and salt in food processor. Add chilled butter and cut in using on/off pulses until mixture appears sandy. Add vanilla and process until dough begins to come together. Press dough evenly into prepared pan. Bake crust 25-28 minutes, or until golden brown.

Topping:

5 large eggs, at room
temperature
2 cups sugar
1 cup strained fresh lemon juice

3 tablespoons cake flour
3 tablespoons grated lemon
peel

Reduce oven to 325°.

While crust is baking, prepare topping. Whisk eggs and sugar in medium bowl to blend. Whisk in lemon juice, then flour. Strain into another bowl. Mix in grated lemon peel. Pour filling over hot crust. Bake until sides are set and filling no longer moves in center when pan is shaken, about 22 minutes. Cool on rack. Cover and chill at least 4 hours or overnight. Using foil sides as aid, lift dessert from pan. Fold down foil sides. Cut into 1 inch squares. (If foil does not lift out easily, you can cut squares while bars are still in baking pan.)

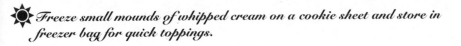

Freeze small mounds of whipped cream on a cookie sheet and store in freezer bag for quick toppings.

CHAMPAGNE DELIGHT

Serves: 4

1 envelope unflavored gelatin
¼ cup cold water
½ cup sugar*
⅓ cup strained fresh orange juice

1½ cups champagne
1 tablespoon crème de cassis (optional)
Strawberries and fresh mint to garnish

Sprinkle gelatin over water in small non-corrosive saucepan. Let stand until soft. Heat over low heat, stirring constantly until gelatin is dissolved, approximately 1 minute. Gradually stir in sugar and orange juice. Heat over low heat, stirring gently until sugar is dissolved, about 2 minutes. (If mixture is stirred too vigorously or whisked, surface of jelly will be foamy.) Remove from heat; cool 5 minutes. Add champagne and, if desired, crème de cassis. Stir gently with a spoon until bubbles subside. Transfer to a small bowl, cover tightly, and refrigerate 4 to 6 hours or until jelly has set. Serve in individual champagne glasses. Garnish with a strawberry and fresh mint.

*Taste your champagne before using. You may wish to adjust the sugar according to the sweetness of the champagne.

APPLE CRISP

Serves: 4

4 cups sliced peeled apples
½-1 cup chopped pecans
1 teaspoon cinnamon
½ teaspoon salt
¼-½ cup water

Juice of ½ lemon
¾ cup sifted flour
1 cup sugar
⅓ cup butter

Preheat oven to 350°.   8 x 8 baking pan

Lightly butter baking pan. In large mixing bowl, combine apples, pecans, cinnamon, salt, water, and lemon juice. Pour mixture into pan. Combine flour and sugar. Cut butter into flour and sugar mixture. Spoon over apples. (Can be frozen at this stage. If frozen, defrost, then cook according to directions.) Bake for 40 minutes.

 Clean oven spills from fruit pie juice by shaking salt over spills. They will burn, and can be easily scraped away.

APPLE DUMPLING DELIGHT

Serves: 6

Pastry Mix:
2¼ cups flour
¾ teaspoon salt

¾ cup shortening
7-8 teaspoons ice water

Sift flour and salt together; cut in shortening with pastry blender. Sprinkle 1 teaspoon of water over mixture. Gently toss with fork. Repeat until all is moistened. Form into a ball. Roll on lightly floured surface. Cut into 6 squares.

Filling:
6 medium tart apples
½ cup sugar
1½ teaspoons cinnamon

1 tablespoon butter or margarine

Preheat oven to 350°.   13 x 9 baking pan

Pare and core apples. Place an apple on each pastry square. Mix sugar and cinnamon together. Fill each apple with cinnamon mixture; dot with butter. Moisten points of pastry; bring points together over tops. Seal well. Place 2 inches apart in pan; place in refrigerator while preparing topping.

Topping:
1 cup sugar
¼ teaspoon cinnamon
2 cups water

4 tablespoons butter or margarine

Mix all ingredients together in saucepan. Boil for 3 minutes. Pour hot syrup around the chilled dumplings; bake 30 to 35 minutes.

Serve warm with whipped cream, if desired.

BANANAS FOSTER

Serves: 4-6

½ cup brown sugar, firmly packed
4 tablespoons butter
1 teaspoon cinnamon
4 bananas, sliced lengthwise

2 jiggers (4 tablespoons) banana liqueur
1 tablespoon dark rum
1 jigger (2 tablespoons) brandy
1 quart vanilla ice cream

Melt sugar and butter in a skillet or chafing dish. Add cinnamon and bananas. Pour liqueur and rum over bananas. Cook until fruit is soft. Add brandy and ignite. Pour over vanilla ice cream; serve immediately.

243

BANANA PUDDING

Serves: 8

1 cup sugar
2 tablespoons flour
3 tablespoons margarine
Pinch of salt
1 teaspoon vanilla extract
2 eggs, well beaten

2 cups milk
4 bananas, sliced
½ box vanilla wafers
Whipped cream
Wafer crumbs

Mix sugar, flour, margarine, salt, vanilla, eggs, and milk together. Cook in a double boiler 5 to 10 minutes, or until thick. Layer sliced bananas and vanilla wafers in serving dish or individual serving bowls. Pour sauce on top; repeat layers. Sprinkle wafer crumbs on top of pudding. Chill thoroughly and serve with whipped cream.

FLUFFY BLUEBERRY DELIGHT

Serves: 14

2 cups blueberries
¼ cup sugar
2 large egg whites

2 tablespoons lemon juice
⅛ teaspoon vanilla extract

9 x 9 metal pan

Freeze blueberries, until hard, in a single layer in pan. Using metal blade with food processor, process fruit until snowy textured, stopping machine occasionally to scrape down sides with a rubber spatula. Place fruit in large mixing bowl; add remaining ingredients. Mix until mixture triples in volume, about 10 minutes, occasionally scraping down sides of bowl. Serve at once or freeze up to 2 hours before serving.

May substitute fresh fruit of choice, cutting fruit into chunks before freezing.

A good "light and healthy" dessert!

 Toast coconut in the microwave. Watch closely as it browns. Spread 1/2 cup coconut in a pie plate and cook for 3 to 4 minutes, stirring every 30 seconds after 2 minutes.

BLUEBERRY PEACH TRIFLE

Serves: 15-20

1 (14 oz.) can sweetened
   condensed milk
1½ cups cold water
2 teaspoons grated lemon rind
1 (3½ oz.) package instant
   vanilla pudding and pie filling
   mix
1 pint whipping cream, whipped

4 cups cubed pound cake
1 pound ripe, fresh peaches,
   or 1 (29 oz.) can of sliced
   peaches in light juice, well
   drained
2 cups fresh or dry packed
   frozen blueberries, thawed,
   rinsed well, and drained

Core, pare, and chop fresh peaches or chop sliced peaches; set aside. In a large bowl, combine condensed milk, water, and lemon rind; mix well. Add pudding mix; beat until well blended. Chill 5 minutes. Fold in whipped cream. Spoon 2 cups pudding mixture into serving dish; top with half the cake cubes, all the peaches, half the remaining pudding mixture, the remaining cake cubes, then the blueberries. Spread the remaining pudding mixture to within one inch of edge of bowl. Chill at least 4 hours. Garnish as desired. Refrigerate leftovers.

Flavor peaches with amaretto; drain well, then use as directed.

CARAMEL CREAM TART

Serves 8-12

1 (8 oz.) package cream cheese
½ cup sugar
1 teaspoon vanilla extract
¼ teaspoon almond extract

6 eggs
2 cups light cream
⅔ cup caramel ice cream
   topping

Preheat oven to 350°.   9 inch round pie pan (1½ inch sides.)

Butter pie pan. With a mixer, beat cream cheese until smooth; add sugar, vanilla, and almond extract. Add eggs, one at a time, beating after each addition; blend in cream. Pour caramel topping into pie pan. Carefully (very slowly) pour cheese mixture on top of caramel topping. Set pan into larger pan of boiling water. Have water within ¾ inch of top of pan. Bake 45 to 50 minutes or until center is set when touched (do not insert knife into center to test.) Cool on a rack; loosen from pan around edges with a knife. Invert on a rimmed cake plate. Spoon remaining sauce over top, cover loosely, and chill. Serve with a coffee liqueur and demitasse.

BREAD PUDDING

Serves: 8-10

¼ pound butter, softened
1 cup plus 2 tablespoons sugar
2 (12 oz.) cans evaporated milk
3 eggs
3 teaspoons vanilla
1 teaspoon cinnamon
¾ teaspoon nutmeg
½ teaspoon salt

¼ teaspoon cream of tartar
¼ teaspoon ginger
4 cups thick bread, torn into
small pieces and toasted
or
4 cups stale cake

Preheat oven to 450°.   13 x 9 baking pan

Prepare baking dish. Beat butter and sugar 5 minutes. Add milk, eggs, vanilla, cinnamon, nutmeg, salt, cream of tartar and ginger. Beat until blended. Line bottom of pan with toasted bread or cake. Pour milk mixture over bread and let sit for 2 hours, patting down and mixing occasionally. Bake for 20-25 minutes until browned. While pudding is baking, prepare Praline Sauce on page 270. Spoon warm praline sauce over warm bread pudding.

CHERRY TORTE

Serves: 6-8

1 cup unsifted flour
2 tablespoons powdered sugar
7 tablespoons margarine
1 (3 oz.) package cream
cheese, softened

½ cup powdered sugar
1 teaspoon vanilla extract
1 cup whipping cream, stiffly
beaten
1 (16 oz.) can cherry pie filling

Preheat oven to 325°.   9 inch pie pan

Mix flour and 2 tablespoons powdered sugar. Melt margarine; pour over dry ingredients. Pat into pie pan, prick, and bake 15-20 minutes. Mix cream cheese, ½ cup powdered sugar, and vanilla. Fold in whipped cream. Spread on crust; top with pie filling. Chill until ready to serve.

CHOCOLATE TRUFFLE DESSERT

Serves: 10-12

1 (8½ oz.) package chocolate
  wafer cookies, crushed
  (about 2 cups)
6 tablespoons butter or
  margarine, melted
1 pound semisweet chocolate
½ cup sugar
2 eggs

4 eggs, separated
2 cups whipping cream
6 tablespoons powdered sugar
2 cups whipping cream
¼ cup powdered sugar
½ teaspoon vanilla extract
  Chocolate leaves

10 inch springform pan

Combine cookie crumbs and butter in a bowl, mixing well. Press on bottom and 2½ inches up sides of ungreased springform pan. Chill 30 minutes.

Place semi-sweet chocolate in top of a double boiler; bring water to a boil. Reduce heat to low, stirring occasionally, until chocolate melts. Remove from heat; stir in ½ cup sugar. Cool to lukewarm. Beat 2 eggs and 4 egg yolks slightly. Gradually stir about one-fourth of warm chocolate mixture into beaten eggs; add to remaining warm mixture, stirring constantly. Beat whipping cream in a large mixing bowl until foamy. Gradually add powdered sugar, beating until soft peaks form. Beat 4 egg whites (at room temperature) at high speed of an electric mixer until stiff peaks form. Fold whipped cream and egg whites into chocolate mixture. Spoon mixture into prepared crust. Cover and chill at least 8 hours. Beat whipping cream in a medium mixing bowl until foamy; gradually add ¼ cup powdered sugar, beating until stiff peaks form. Stir in vanilla. Set aside ½ cup whipped cream, spread remaining whipped cream over chocolate filling. Remove sides of pan and pipe reserved ½ cup whipped cream around edge. Garnish with chocolate leaves. Store in refrigerator.

Chocolate Leaves:

8 (1 oz.) squares semi-sweet
  chocolate
1 tablespoon shortening

Nonpoisonous leaves
(mint or rose)

Wash leaves; pat dry with paper towels. Melt chocolate and shortening over hot water in double boiler; let cool slightly.

Using a small spatula, spread a ⅛-inch layer of melted chocolate on the back of each leaf, spreading to the edges. Place chocolate-coated leaves on wax paper-lined baking sheet, chocolate side up; freeze until chocolate is firm, about 10 minutes. Grasp leaf at stem end; carefully peel leaf from chocolate. Chill chocolate leaves until ready to use. (Handle carefully since chocolate leaves are thin and will melt quickly from the heat of your hand.)

DEATH BY CHOCOLATE

Serves: 15-18

1 (22 oz.) box brownie mix
½ cup Kahlúa
2 (3.5 oz.) packages chocolate
  mousse pudding mix
1 (16 oz.) container whipped
  topping

8 Heath or Skor candy bars,
  crushed
1 cup chopped pecans

Preheat oven to 350°.   13 x 9 pan

Grease pan. Prepare brownie mix according to the directions on the box. When the brownies are slightly cooled, poke holes in the top with a fork; pour Kahlúa over the top. When the brownies have thoroughly soaked in the Kahlúa, crumble into pieces; return to pan. Prepare mousse mix according to package directions. Assemble in layers, smoothly spread ½ chocolate mousse over crumbled brownies, then smoothly spread ½ whipped topping over mousse. Sprinkle ½ crushed candy bars over whipped topping, repeat layers, and top with pecans.

CARAMEL ICE CREAM DESSERT

Serves: 10-12

2 cups flour
½ cup oatmeal
½ cup brown sugar, firmly
  packed
1 cup chopped nuts

1 cup butter, melted
2 (12 oz.) jars caramel ice
  cream topping
½ gallon vanilla ice cream,
  softened

Preheat oven to 400°.   13 x 9 pan

Mix flour, oatmeal, and brown sugar. Add nuts and butter; mix well. Spread mixture on large cookie sheet. Bake for 15 minutes, crumble while hot. Cool mixture and sprinkle half of crumbs on bottom of pan. Drizzle 1 jar of caramel topping over crumbs. Spread ice cream on top of caramel. Sprinkle remaining crumbs; drizzle other jar of topping over crumbs. Freeze until ready to serve. Let soften a little and cut into squares.

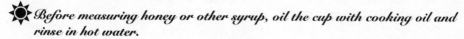

Before measuring honey or other syrup, oil the cup with cooking oil and rinse in hot water.

HONEY VANILLA ICE CREAM

Yield: 1 gallon

4 eggs, separated
2 cups sugar
1 (13 oz.) can evaporated milk
1 pint whipping cream

2 pints half-and-half
¾ cup honey
1 teaspoon vanilla extract

Beat egg whites until stiff; slowly add sugar to whites. In another bowl, beat yolks until foamy; add to whites. Blend in evaporated milk, whipping cream, half-and-half, honey, and vanilla. Pour into ice cream freezer and follow freezer instructions.

Fresh fruit of choice or other flavorings may be added to basic recipe.

ORANGE FREEZE

Serves: 8

1 (6 oz.) package orange
  gelatin
2 cups boiling water
1 pint orange sherbet
1 (8 oz.) carton refrigerated
  whipped topping

1 tablespoon sugar
1 (11 oz.) can mandarin
  oranges
1 (20 oz.) can crushed
  pineapple

13 x 9 glass dish

Dissolve gelatin in water. Refrigerate 10 minutes. Mix sherbet and topping. Fold into gelatin mixture with other ingredients. (To prevent breaking, fold in oranges last.) Put in dish or molds. Freeze. Before serving, place in refrigerator to soften slightly. Cut and serve. Can be molded in paper muffin cups. Just peel off the paper and serve.

 Cream will whip faster and better if you will first chill the cream, bowl, and beaters well.

LEMON CHEESE DESSERT

Serves: 12-15

1 (3.4 oz.) package lemon
  gelatin
1 cup boiling water
2-5 tablespoons fresh lemon
  juice, to taste
1 (8 oz.) package cream cheese

1 cup sugar
1 teaspoon vanilla extract
1 (13 oz.) can chilled
  evaporated milk
1 box vanilla wafers

9 x 9 glass dish

Mix gelatin, water, and lemon juice. Set aside. In a separate bowl, mix cream cheese, sugar, and vanilla until creamy; pour into gelatin mixture. Using another bowl, whip chilled milk until fluffy. Fold in gelatin and cheese mixture. Crush vanilla wafers, reserving ½ cup for topping, and line dish. Pour dessert into dish. Sprinkle top of dessert with remaining crushed wafers. Chill until set. Keep refrigerated.

NILLA WAFERS DESSERT

Serves: 8

1 box vanilla wafers
2 bananas
1 (20 oz.) can crushed
  pineapple, well-drained
1 (10 oz.) can sweetened
  condensed milk

2 tablespoons lemon juice
Whipped cream
Maraschino cherries

9 inch pie pan

Line bottom and sides of pie pan with wafers. Slice bananas and place on top of wafers; reserve a few slices for decoration. Mix pineapple with milk and lemon juice. Pour mixture on top of wafers and bananas. Top with whipped cream. Decorate with banana slices and cherries. Chill for 2 hours. Keep refrigerated.

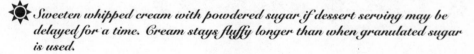 *Sweeten whipped cream with powdered sugar if dessert serving may be delayed for a time. Cream stays fluffy longer than when granulated sugar is used.*

RAINBOW ICE CREAM DESSERT

Serves: 15

4 (3 oz.) boxes of gelatin, 4 different colors and flavors
½ gallon vanilla ice cream, softened, divided

4 cups water, divided

10 inch tube pan

Dissolve 1 box gelatin in 1 cup of boiling water. Mix in 2 cups of ice cream. Pour mixture into tube pan. Freeze until the first layer has set, approximately 45 minutes. Repeat procedure to make the next layer. When gelatin is sticky, spoon second layer over first layer. Continue these steps for layers three and four. Serve as a salad on a hot summer day or as dessert anytime.

TEXAS PEACH COBBLER

Serves: 8-10

½ cup margarine
1 cup self-rising flour
1 cup sugar

1 cup buttermilk
1 teaspoon vanilla extract
2-3 cups fresh or frozen peaches

Preheat oven to 350°.

3 quart casserole dish

Place margarine in deep dish; place in oven. Meanwhile, mix flour, sugar, buttermilk, and vanilla until smooth. When margarine is melted and bubbling, remove from oven and pour flour mixture into casserole. Spoon fruit on top of flour mixture; return to oven. Bake 25-30 minutes or until crust is brown. Also good with other fruits.

A deep dish is essential.

GOURMET PEARS

Serves: 4

4 medium pears
¼ teaspoon mace or nutmeg
½ cup brown sugar, firmly packed

¼ cup butter, cut up
¼ cup dark rum
½ cup whipping cream

Preheat oven to 375°.

8 inch pie pan

Peel pears, leaving stem. Slice bottoms so they sit evenly in pan. Mix sugar and mace; sprinkle over pears. Dot with butter; add rum. Bake for 45 minutes. Remove pears; place in serving dish. Boil liquid, add cream, and boil until thickened. Spoon over pears; serve. Easy!

PISTACHIO DESSERT

Serves: 15-20

40 Ritz crackers, crushed
1 teaspoon sugar
¼ cup margarine
2 (3 oz.) boxes instant
pistachio pudding mix

1½ cups milk
1 quart softened vanilla ice
cream
3 frozen Heath candy bars

Preheat oven to 300°.   13 x 9 baking pan

Combine crushed crackers with sugar and margarine. Press into pan and bake for 10 minutes. Mix pudding and milk; set in refrigerator for 10 minutes to chill. Add softened ice cream; pour over crust and freeze for 2 hours. Remove from freezer and let stand 30 minutes. Top with whipped topping and grated Heath bars prior to serving.

FRUIT PIZZA

Serves: 8

1 (18 oz.) package refrigerated
slice-and-bake sugar cookie
dough
1 (8 oz.) package cream cheese
⅓ cup sugar
½ teaspoon vanilla extract
Banana slices (for rim)

Blueberries, pineapple, orange
sections, strawberries,
peaches (whatever fruit you
enjoy)
¼ cup apricot preserves
1 tablespoon water

Preheat oven to 375°.   14 inch pizza pan

Cut sugar cookie dough into ⅛ inch slices. Line pizza pan with slices overlapping. Bake dough for 12 minutes. Cool cookie crust. Blend cheese, sugar, and vanilla. Spread cream cheese mixture over cookie crust. Arrange fruit in circular pattern over cream cheese starting with bananas on outside rim. Mix apricot preserves with water; spread preserves over fruit. Be sure to cover all fruit so no discoloration occurs. Refrigerate until chilled.

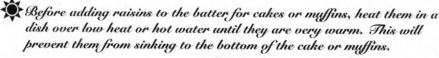

Before adding raisins to the batter for cakes or muffins, heat them in a dish over low heat or hot water until they are very warm. This will prevent them from sinking to the bottom of the cake or muffins.

STRAWBERRIES WITH ORANGE CREAM
Serves: 8-10

2 quarts fresh strawberries
1¼ cups sugar, divided
4 teaspoons grated orange rind

1 cup orange juice
1 (12 oz.) carton Lite Cool Whip

Wash and hull berries; cut into halves or quarters. Combine with ¼ cup sugar. Cover and chill. Combine 1 cup sugar, orange rind, and orange juice in a small saucepan. Bring to a boil; stir only until the sugar dissolves. Simmer 15 minutes without stirring. Cool completely. Prior to serving, fold orange syrup into Lite Cool Whip until well blended. Serve over berries.

COFFEE ICE CREAM PIE
Serves: 6-8

Crust:

2 tablespoons margarine
2 (1 oz.) squares unsweetened baking chocolate

2 tablespoons hot milk
⅔ cup sifted powdered sugar
1½ cups shredded coconut

9 inch pie pan

Grease pie pan with non-stick vegetable spray. Melt margarine and chocolate over low heat. Stir well. In a small bowl, combine hot milk with powdered sugar and add to chocolate mixture. Stir in coconut and press evenly into pie pan. Freeze 2-3 hours or until firm.

Filling:

1 quart vanilla ice cream
2 teaspoons instant coffee crystals

½ cup chopped pecans

Spoon ice cream into a large bowl to soften. Fold the coffee crystals and pecans into ice cream and spread evenly over frozen pie crust. Cover tightly and freeze several hours until firm. Thaw slightly before serving.

CRUNCHY APPLE PIE

Yield: 2 pies

Crust:

3 cups flour
1¼ cups vegetable shortening
1 teaspoon salt

1 egg
1 teaspoon vinegar
5½ tablespoons water

Mix flour, shortening, and salt. Set aside. Stir together egg, vinegar, and water. Add to flour mixture. Roll out 2 pie shells.

Filling:

8-9 Granny Smith apples
1 cup sugar
½ cup flour

1 teaspoon cinnamon
Dash nutmeg

Cut up and peel apples. Mix apples, sugar, flour, cinnamon, and nutmeg together. Put into shells. (This makes one thick pie or two thinner ones.)

Topping:

½ cup butter
½ cup brown sugar, firmly packed

¾ cup flour
Cinnamon

Preheat oven to 350°. 2 (8 inch) pans

Blend butter, brown sugar, and flour with a pastry blender. Sprinkle on top of pies; then sprinkle a little cinnamon on top. Bake for 65-70 minutes. Then put it under the broiler to brown the top.

BUTTERMILK COCONUT PIE

Serves: 8-10

1½ cups sugar
2 tablespoons flour
½ cup butter, melted
3 eggs, well beaten
½ cup buttermilk

1 teaspoon vanilla extract
1 heaping cup or 1 (3¼ oz.) can coconut
1 (9 inch) pastry shell, unbaked

Preheat oven to 350°. 9 inch pie pan

Mix all the ingredients together; pour in an unbaked pie shell. Place on a cookie sheet; bake in oven for 1 hour, or until firm. If the crust starts to brown too rapidly, tent it with foil.

BLUEBERRY CREAM CHEESE PIE
Serves: 8

1 (8 oz.) package cream
  cheese, softened
¾ cup sugar, divided
¼ cup sour cream
½ teaspoon vanilla extract
1 (9 inch) pastry shell, baked
2½ cups fresh blueberries, divided
½ cup water

2 tablespoons cornstarch
1 tablespoon lemon juice
Sour cream
2 tablespoons powdered sugar
Dash of ground cinnamon
Blueberries
Grated nutmeg

9 inch pie pan

Combine cream cheese, ¼ cup sugar, sour cream, and vanilla. Beat at medium speed of an electric mixer 1 minute or until smooth. Spread mixture in baked pastry shell; cover and chill for 1 hour.

Place 1 cup blueberries in a small saucepan; mash with a potato masher. Add water; bring to a boil. Reduce heat and simmer 3 minutes. Remove from heat. Strain berry mixture. If necessary, add water to juice to make 1 cup; return to saucepan. Combine ½ cup sugar and cornstarch; stir into berry liquid. Bring to a boil, and boil 1 minute, stirring constantly. Stir in lemon juice; cool.

Place 1½ cups blueberries over cream cheese mixture; pour berry liquid glaze over top. Chill 2-3 hours. To serve, dollop each slice with sour cream. Combine powdered sugar and cinnamon; sprinkle mixture over sour cream. Garnish with blueberries and freshly grated nutmeg.

TEXAS PECAN PIE
Serves: 8

1 cup white corn syrup
1 cup dark brown sugar, firmly
  packed
⅓ teaspoon salt
⅓ cup margarine, melted

1 teaspoon vanilla extract
3 eggs, slightly beaten
1 heaping cup pecan halves
1 (9 inch) pie shell, unbaked

Preheat oven to 350°.
9 inch pie pan

Combine syrup, sugar, salt, margarine, and vanilla; mix well. Add eggs. Pour into pie shell. Sprinkle pecans over all. If time allows, arrange pecans neatly in circles on top of pie. Bake for 45-55 minutes. (Sharp knife inserted in pie should come out clean).

CHOCOLATE PIE

Serves: 8

½ cup butter
¾ cup sugar
4 tablespoons cocoa
1 teaspoon vanilla extract

2 eggs
1 (8 inch) pie shell, baked
Whipped cream
Grated chocolate

8 inch pie pan

Cream butter and sugar very well. Blend in cocoa and vanilla. Add eggs, one at a time. Beat at medium speed for 5 minutes after each egg. Pour into pie shell; chill for 1-2 hours. Top with whipped cream and grated chocolate.

FT. BROWN PIE

Serves: 6-8

Chocolate Pecan Crust:
2 cups chocolate cookie crumbs
¾ cup finely chopped pecans

½ cup margarine, melted

Preheat oven to 350°.

9 inch pie pan

Combine ingredients for crust. Press into pan covering bottom and sides. Bake 10 minutes; set aside to cool.

Filling:
1 envelope unflavored gelatin
¼ cup cold water
2 cups whipping cream, divided
6 ounces semi-sweet chocolate
  chips

2 eggs
1 teaspoon vanilla extract
22 caramels (about 1 cup)
2 tablespoons margarine

In a small saucepan, sprinkle gelatin over water; let stand 1 minute. Stir over low heat until completely dissolved. Stir in 1 cup cream. Bring to a boil. In blender, process gelatin mixture with chocolate chips. Add ½ cup cream, eggs, and vanilla. Process until chips are melted and mixture is well blended. Pour into bowl and chill until thickened. In a small saucepan, combine caramels, scant ¼ cup cream, and margarine. Simmer, stirring frequently until caramels are melted. Pour into cooled crust. Let stand for 10 minutes. With whisk, beat thickened gelatin mixture until smooth. Pour gently over caramel layer in crust. Chill until firm, 4-6 hours or overnight. Garnish with remaining cream, whipped.

GRAPEFRUIT PIE

Serves: 6-8

1 cup sugar
1¾ cups water
2 tablespoons cornstarch
⅛ teaspoon salt
1 (3 oz.) package strawberry
gelatin

3 or 4 grapefruit, sectioned and
drained
1 (9 inch) pie crust, baked
Whipped topping

Cook sugar, water, cornstarch, and salt until thick. Add gelatin. Stir until dissolved. Add grapefruit. Put filling in pie crust. Top with whipped topping.

GRASSHOPPER PIES

Serves: 12

Pie Filling:
30 large marshmallows
½ cup milk
¼ cup green crème de menthe

1¼ cups white crème de cacao
1¼ cups heavy whipping cream

Muffin tin with paper liners

In the top of a double boiler over hot, not simmering, water, melt the marshmallows with the milk, stirring frequently until smooth. Remove the top part of the double boiler from the bottom and cool the marshmallow mixture for 15 to 20 minutes, or until tepid. Stir in the liqueurs, or place the marshmallows and milk in a 3 quart microwave-safe glass bowl. Microwave on high power for 2 to 3 minutes. Remove the bowl from the microwave and whisk the mixture until smooth. Let the mixture cool for 15 minutes. Stir in the liqueurs. In a chilled, large bowl, using a hand-held electric mixer set at medium-high speed, beat the cream until soft peaks begin to form. Gently fold the whipped cream into the marshmallow mixture using a large rubber spatula.

Crust:
1¼ cups chocolate wafer cookie
crumbs

1 tablespoon unsalted butter,
melted

In a small bowl, stir together the cookie crumbs and butter. Sprinkle 1½ teaspoons of the cookie crumbs into the bottom of each of the lined cups. Place 2 tablespoons of pie filling over crumbs in each cup. Sprinkle 2 teaspoons of cookie crumbs over the pie filling. Divide the remaining pie filling among the cups. Freeze the Grasshopper Pies for 8 hours or overnight. To serve, remove the pie cups from the pan and carefully peel off the paper liners. Garnish with additional whipped cream and cookie crumbs, if desired.

LEMON ALASKA PIE
Serves: 8

Crust:
- 1 cup finely chopped pecans
- 4 tablespoons unsalted butter, softened
- ¼ cup sugar
- 1 tablespoon flour

Preheat oven to 450°.   10 inch springform pan

Line springform pan with aluminum foil. Combine crust ingredients. Press mixture into bottom of pan; bake for 10 minutes or until lightly browned. Cool on a rack.

Filling:
- 2 pints vanilla ice cream, softened, divided

Spread crust with 1 pint ice cream. Freeze for 30 minutes, or until ice cream is hard.

Sauce:
- 1½ cups sugar
- 9 tablespoons unsalted butter, melted
- ½ cup lemon juice
- 1 tablespoon grated lemon rind
- ¼ teaspoon salt
- 3 large eggs
- 3 egg yolks

Mix all sauce ingredients together except eggs. Beat in eggs and egg yolks until all is well combined. Cook sauce in top of double boiler over simmering water, stirring for 10-20 minutes or until thickened. Put sauce in a bowl, cool, cover loosely, and chill. Spread half of the chilled sauce mixture over ice cream; freeze 30 minutes. Spread the second pint of softened ice cream over sauce; freeze for 30 minutes again. Spread remaining sauce over ice cream; freeze for 8 hours or overnight. Remove pan sides; let dessert soften at room temperature for 5 minutes. Cut into 16 wedges; peel off foil. Arrange wedges, well separated, on a large baking sheet; freeze.

(Continued on next page)

(Lemon Alaska Pie, continued from previous page)

Meringue:
  6 egg whites
  ¼ teaspoon cream of tartar
  ¼ teaspoon salt
  ½ cup sugar

Just before serving, preheat broiler and prepare meringue. Beat egg whites, cream of tartar, and salt until they hold soft peaks. Gradually add sugar, beating until meringue is stiff and shiny. Spread meringue over tops and sides of ice cream wedges. Place 6 inches from broiler heat for 2 minutes or until meringue is browned. Serve immediately.

You can freeze ahead and take out pieces, one at a time, to cover with meringue and bake.

STRAWBERRY PIE

Serves: 6

Crust:
  ½ cup margarine
  2 tablespoons sugar
  1 cup flour

Preheat oven to 375°.   9 inch pie pan

Mix crust ingredients together and press into pie plate. Bake for 10-15 minutes, until golden brown.

Filling:
  1-2 pints fresh strawberries
  1 cup sugar
  4 tablespoons cornstarch
  ⅓ cup water
  Juice of ½ lemon
  Dash of salt
  1 (10 oz.) package frozen strawberries
  Whipped cream

Fill bottom of baked pie shell with fresh, whole strawberries that have been washed, hulled, and drained on paper towels. (If they aren't dried well, the crust will be soggy.) Cook remaining ingredients, except whipped cream, together until thick and clear, stirring constantly. Cool. Pour over fresh strawberries in shell. Top with sweetened whipped cream and garnish with 1 fresh strawberry on each piece.

PRALINE PIE

Serves: 6-8

½ cup margarine
¼ cup brown sugar, firmly
   packed
1 cup flour

½ cup chopped pecans
½ gallon vanilla ice cream
Butterscotch sauce

Preheat oven to 400°.   9 inch pie pan and cookie sheet

Melt margarine; mix with brown sugar, flour, and pecans. Mash on cookie sheet; bake for 7-10 minutes until lightly browned. While warm and crumbly, take ⅔ of mixture; press into the bottom of pie pan. Fill the rest of the pie pan with ice cream; sprinkle the rest of the crumbs over the top. Freeze. When ready to serve, cut into pieces, and drizzle butterscotch sauce over the top of each portion.

PUMPKIN PIE

Serves: 8

2 eggs
1 (16 oz.) can pumpkin
¾ cup sugar
½ teaspoon salt
1 teaspoon cinnamon
½ teaspoon ginger

¼ teaspoon cloves
1 teaspoon nutmeg
2 tablespoons bourbon
1⅔ cups evaporated milk
1 (9 inch) pie shell, unbaked
Whipped cream

Preheat oven to 425°.   9 inch pie pan

Slightly beat eggs. Mix eggs with the pumpkin, sugar, salt, cinnamon, ginger, cloves, nutmeg, bourbon, and evaporated milk. Pour mixture into pie shell. Bake for 15 minutes. Reduce temperature to 350° and continue baking for 45 minutes. Cool; top with whipped cream before serving.

Mas, Mas, Mas!

MAS, MAS, MAS!

Nursing Homes
The nursing home program has been a longstanding volunteer project. Our sustainer volunteers entertain residents of nursing homes with holiday parties and regular bingo games. These volunteers have provided a much needed service by giving the residents of nursing homes love and attention to help fill their lives with happiness.

Hospice
Our goal for the Hospice program is to provide support for terminally ill patients and their families. Our volunteers help the children and family members of the terminally ill by becoming their friend, meeting some of their needs, and helping them cope during a stressful situation.

 These recipes are pictured on the previous page.

CRANBERRY CHUTNEY

Yield: 4 cups

½ cup apricot preserves
½ cup cider vinegar
½ cup brown sugar, firmly
  packed
¾ teaspoon curry
¾ teaspoon ground ginger
  Cheesecloth bag
1 (3 inch) cinnamon stick
6 whole cloves

1½ cups water
1 lemon
1 firm pear, peeled and diced
1 Granny Smith apple, peeled
  and diced
3 cups cranberries
½ cup raisins
½ cup walnuts

Seed, chop, and blanch lemon in boiling water for 2 minutes, set aside. In a saucepan combine preserves, vinegar, sugar, curry, ginger, and cheesecloth bag containing cinnamon stick and cloves. Add water. Bring to a boil until sugar dissolves. Add lemon, (drained) pear, and apple. Simmer 10 minutes. Add cranberries and raisins, stirring occasionally. Remove from heat. Stir in walnuts. Discard cheesecloth bag.

TROPICAL CHUTNEY

Yield: 5-6 pint jars

6 cups vinegar
4 cups granulated sugar
2 cups brown sugar, firmly
  packed
3 teaspoons salt
3 teaspoons ground cloves
3 teaspoons allspice

1 cup crystallized ginger
4 cups pineapple chunks
4 cups papaya chunks
4 cups mango chunks
1½ cups chopped onion
3 cloves garlic, chopped
3 cups white raisins

In a large pot, bring sugar and vinegar to a boil and cook for 10 minutes, until sugar is dissolved; add mangos, spices, onions, garlic, raisins, and ginger, and boil 10 minutes; add rest of fruit and simmer 20-30 minutes. Remove from heat, and with a slotted spoon, divide fruit into warm sterile jars. Seal tightly with paraffin or self-seal lids.

 When a drain is clogged with grease, pour 1 cup of salt and 1 cup of baking soda followed by a kettle of boiling water into drain.

CREAMY FRUIT DIP

Yield: 4 cups

1 cup powdered sugar
1 (12 oz.) package cream cheese, softened
1 (7 oz.) jar marshmallow cream

1 (8 oz.) carton sour cream
2 teaspoons vanilla extract
2 teaspoons almond extract
2 teaspoons cinnamon

Mix all ingredients together. Keep refrigerated.

KAHLÚA AND CREAM FRUIT DIP

Yield: 4 cups

1 (8 oz.) package cream cheese
1 (8 oz.) carton whipped topping
¾ cup light brown sugar, firmly packed

⅓ cup Kahlúa liqueur
1 cup sour cream

Blend together cream cheese and whipped topping. Add sugar and Kahlúa. Mix well. Add sour cream. Refrigerate one or two days before serving with fresh fruit.

SOUTH OF THE BORDER SUBS

Serves: 6-8

2 cups grated Cheddar cheese
1 (7½ oz.) can Salsa Verde
1 (4 oz.) can chopped black olives
⅓ cup olive oil

3 green onions, chopped
2 jalapeños, seeded and diced (optional)
1 loaf French bread, cut into ½ inch slices

Preheat broiler.

In a medium bowl, mix together cheese, Salsa Verde, black olives, olive oil, onions, and jalapeños. Arrange bread slices on broiler pan and spread each piece with cheese mixture. Broil until bubbly, 1 or 2 minutes. Serve immediately.

JALAPEÑO JELLY

Yield: 5-6 half pints

½ cup chopped fresh jalapeños (6-8)
1 red bell pepper, seeded and finely chopped
1 green or yellow bell pepper, seeded and finely chopped

5 cups sugar
1¼ cups wine or white vinegar
6 ounces liquid fruit pectin
Paraffin

Sterilize 6 half pint jars.

Place the peppers in a large saucepan. Add sugar and vinegar. Bring to a boil; boil for 5 minutes uncovered. Remove saucepan from heat. Skim off any foam that comes to the top of the liquid mixture as it cools for 20 minutes. Add pectin. Mix well; bring to a full rolling boil a second time. Remove from heat; stir well. Skim off any foam; carefully pour hot mixture into sterilized jelly jars. Seal the jars with melted paraffin while jelly is hot. Do not place tops on jars until paraffin has sealed.

As an appetizer, put jelly on top of any size cream cheese and serve with crackers. Serve with lamb or poultry; mix with mayonnaise as a sauce for cold, boiled shrimp. Serve as a sauce for eggrolls.

UNCOOKED ORANGE MARMALADE

Yield: 4-5 pints

2 oranges
1 lemon
1 pound dried apricots, soaked overnight in hot water to cover

1 (20 oz.) can crushed pineapple
4-5 pounds granulated sugar

Wash oranges and lemon well; remove seeds and as much white pulp as possible. Grind fruit, peel and all (drain apricots prior to grinding). Mix fruit with undrained pineapple; add equal amount of sugar. Place in refrigerator; stir daily for 3 days. Marmalade may be put in pint jars for gifts. Will keep indefinitely in refrigerator.

MEAT MARINADE

¼ cup vegetable or olive oil
1 tablespoon wine vinegar
1 teaspoon onion powder
1 teaspoon garlic powder
1 teaspoon cayenne pepper
1 teaspoon Louisiana hot sauce
6 tablespoons fresh lemon juice

1 tablespoon salt
1 teaspoon pepper
2 tablespoons Creole mustard
1 tablespoon prepared horseradish
3 cans of beer

In a large mixing bowl, combine ingredients and mix well. Place meat in marinade and refrigerate for 1-3 days. Large zip bags work well to hold meat and marinade. Grill meat on barbeque pit.

Suggested meats: steak, ribs, fajitas.

PERFECT MARINADE

Yield: 1 quart

1½ cups peanut oil or vegetable oil
¾ cup soy sauce
2 tablespoons Worcestershire sauce
2 tablespoons dry mustard
2 tablespoons salt

1 tablespoon pepper, freshly ground
½ teaspoon garlic salt
1 cup dry red wine
2 teaspoons dried parsley flakes
⅓ cup lemon juice

Combine all ingredients in a 1 quart jar. Seal tightly and shake vigorously. An excellent marinade for steak, brisket, beef cubes, or kabobs.

BARBEQUE SAUCE

Serves: 4-6

1 (14 oz.) bottle of ketchup
2 ounces Worcestershire sauce
¾ cup vinegar
½ teaspoon Tabasco sauce
½ cup light brown sugar, firmly packed

½ cup butter or margarine
1 tablespoon salt
1 tablespoon black pepper
1 tablespoon garlic juice
2 tablespoons liquid smoke

In a large saucepan, mix all ingredients. Bring to a boil.

SOUTH TEXAS BARBEQUE SAUCE

1 large onion, diced
2 tablespoons oil
1 cup vinegar
3 teaspoons prepared mustard

⅓ cup Worcestershire sauce
1 cup beer
Salt and pepper to taste

In a medium saucepan, sauté onion in oil. Add vinegar, mustard, and Worcestershire sauce. Mix well and simmer for a few minutes. Add beer, salt, and pepper.

BROILED RUBY RED GRAPEFRUIT

Serves: 1

1 Texas red grapefruit

Cut a grapefruit in half; loosen each section with a sharp knife. Cover with desired topping and place 3 to 5 inches from heat source. Broil until bubbly, about 3 to 6 minutes.

Several topping ideas are brown sugar and cinnamon; maple or fruit-flavored syrup; melted butter mixed with chopped nuts, brown sugar and coconut; brown sugar mixed with your favorite liqueur or flavoring. Try brandy, sherry, kirsch, Grand Marnier, or amaretto.

Broiled grapefruit is quick and easy. It is delicious as a breakfast treat, a light appetizer before dinner, or even a refreshing dessert.

CORN SALSA

Yield: 4 cups

4 ears fresh corn
1 cup diced zucchini squash
3 tablespoons olive oil
3 medium tomatoes, diced
1 serrano chili, seeded and
  minced

½ cup chopped cilantro
1 medium onion, diced
2 tablespoons lime juice
2 teaspoons liquid smoke
½ teaspoon cumin
1 (8 oz.) can tomato sauce

Boil corn for 2-3 minutes; cut off cob. In a skillet, sauté squash in olive oil until squash is bright green, but still crisp. Remove from heat; add remaining ingredients. Mix well and refrigerate. Serve cold.

Excellent served with chips, broiled or grilled fish or chicken.

ORANGE BUTTER

¼ cup orange juice, frozen and
   undiluted
10 tablespoons butter or
   margarine

1 (16 oz.) box powdered sugar

Cream ingredients together. Put in covered container and refrigerate.

EGGPLANT PASTA SAUCE

Yield: 2 quarts

⅓ cup olive oil
2-3 garlic cloves, minced
1 medium (1 lb.) eggplant,
   unpeeled and chopped
2 green bell peppers, diced
3 cups tomatoes, peeled and
   chopped

¾ cup sliced black olives
4 tablespoons capers
1 teaspoon crushed oregano
½ teaspoon crushed basil
12 ounces tomato paste
2 cups dry white wine
   Salt and pepper to taste

In a large skillet, sauté garlic in oil. Add eggplant, pepper, tomatoes, olives, and capers. Stir well to combine with oil. Add remaining ingredients. Cover and simmer for 1 hour. Stir occasionally and add more white wine if necessary.

Will keep in refrigerator for three weeks. May be served cold as vegetable or hot over pasta.

GREEN SALSA

Yield: 3 cups

6-8 tomatillos or 3-4 green
   tomatoes
2 jalapeño peppers or 4 serrano
   peppers or 1 tablespoon
   ground chili pequin
1 large onion, quartered

3 cloves garlic, minced
½ teaspoon salt
½ teaspoon pepper
1 teaspoon vinegar
1 tablespoon sugar

Wash, husk, and cut tomatillos in half and cook in boiling water for 5 minutes. Drain; reserving ½ cup water. Place all ingredients in blender with water and mix 1 minute. Can be used fresh or simmered 5 minutes. If you don't choose to use a blender, chop all ingredients fine and mix.

MANGO JICAMA SALSA

Yield: 2¾ cups

2 cups jicama, cut into ¼ inch cubes
1 cup chopped mango
½ cup chopped white onion
2 tablespoons chopped fresh cilantro

1 or 2 chopped fresh jalapeños, seeded
Juice of 1 lemon

Mix all ingredients together. Cover and refrigerate until time to serve. May be served as an appetizer with chips or as a side dish.

RÉMOULADE SAUCE

Yield: 1 pint

1 hard boiled egg, finely chopped
2 shallots, finely chopped
4 garlic buttons, finely chopped
¼ cup fresh spinach, cooked, drained, and finely chopped
2 cups mayonnaise
1 teaspoon Worcestershire sauce

1 teaspoon Creole mustard
1⅓ ounces anchovy paste
⅛ teaspoon Tabasco sauce
2 tablespoons lemon juice
2 tablespoons chopped parsley
1 teaspoon capers

In a large bowl, mix all the ingredients. Chill for several hours to allow flavors to blend.

TEXAS TREAT

Yield: 1 gallon

6 cups Rice Chex cereal
3 cups Cheerios cereal
2 cups pretzels
2 cups salted peanuts

1 (12 oz.) bag M&M's candy
1½ pounds white chocolate or almond bark

Toss cereals, pretzels, peanuts, and candy together. Using a double boiler, melt white chocolate. Toss with cereal mixture. Spread on waxed paper until cool.

TOMATO MUSHROOM SAUCE

Serves: 10-12

4 medium onions, finely chopped
3 small cloves garlic
2 tablespoons olive or salad oil
4 (29 oz.) cans tomatoes
2 (12 oz.) cans tomato paste

1 pound mushrooms, sliced
3 tablespoons sugar
2 bay leaves
4 teaspoons oregano
1 cup chopped fresh parsley

In a large saucepan, over medium heat, cook onions and garlic in oil until onions are limp. Discard garlic, Add remaining ingredients. Reduce heat to low and simmer, covered, for 2 hours. Discard bay leaves.

Can be frozen in 1 pint or larger containers for up to 3 months.

HOT FUDGE SAUCE

Serves: 8

4 (1 oz.) squares unsweetened chocolate
½ cup margarine

⅛ teaspoon salt
2 cups granulated sugar
1⅔ cups evaporated milk

Melt chocolate and margarine in top of double boiler. Slowly stir in salt, sugar, and evaporated milk. Bring to boil stirring constantly. Remove immediately from heat. Cool to room temperature. Put in an airtight container and store in refrigerator. Will keep indefinitely. Reheat again before serving.

PRALINE SAUCE

Serves: 8

4 tablespoons butter
1 cup light brown sugar, firmly packed

1 cup heavy cream
⅛ cup bourbon, 101 proof
⅓ cup chopped nuts

In a small saucepan, over low heat, cream butter and brown sugar; gradually add cream. Stir over heat until mixture comes to a boil. Remove from heat. Add chopped nuts.

Serve warm over bread pudding, ice cream, pound cake; almost anything. Can be made 2-3 days in advance.

CITRUS PECAN TOPPING

Yield: 2 cups

2 Texas red grapefruit
2 Texas oranges
¼ cup sugar
1-1½ tablespoons cornstarch
3 tablespoons water

2 teaspoons margarine
3 tablespoons chopped pecans,
toasted
Dash of salt

Peel and section one grapefruit and one orange over a bowl, reserving juice. Set sections aside. Measure grapefruit juice; squeeze enough juice from other grapefruit to equal 6 tablespoons juice. Measure orange juice; squeeze enough juice from remaining orange to equal 6 tablespoons juice. Set aside.

Combine sugar, cornstarch, and salt in a saucepan; stir in juices and water. Cook over medium heat, stirring constantly, until thickened and bubbly. Add butter and stir until melted. Stir in grapefruit and orange sections and pecans. Serve warm or at room temperature.

A perfect sauce for pound cake, shortcake, bread pudding, or even pancakes.

CUCUMBER SANDWICH SPREAD

Serves: 20

2 medium cucumbers, unpeeled
1 (8 oz.) and 1 (3 oz.) package
cream cheese, softened
¼ cup mayonnaise
3 green onions, finely chopped,
tops and all

2 tablespoons chopped parsley
3 dashes Tabasco sauce
½ teaspoon seasoned pepper
1 teaspoon garlic powder
1 tablespoon chopped chives
Bread, crust removed

Grate cucumber; carefully press out moisture with paper towels. Mix all ingredients well, using enough mayonnaise to moisten so that filling will spread easily. Spread on bread. Refrigerate or freeze.

 Loosen grime from can openers by brushing with a toothbrush. To clean blades, run a paper towel through the cutting process.

GRANOLA

Yield: approximately 2 quarts

4 cups uncooked oats
1½ cups raw or toasted wheat
  germ
1 cup grated coconut
¼ cup nonfat dry milk
2 tablespoons cinnamon
1 tablespoon brown sugar,
  firmly packed

⅓ cup vegetable oil
½ cup honey
1 tablespoon vanilla extract
½ cup raw nuts, seeds, raisins,
  or chopped dates

Preheat oven to 250°.   17 x 14 cookie sheet

In a large bowl, mix dry ingredients. Combine oil, honey, and vanilla in a saucepan and warm; add to dry ingredients. Mix by hand until all particles are coated. Spread mixture on baking sheet. Bake for 1 hour, stirring periodically. When finished toasting, add nuts, seeds, or dried fruits. Cool and store in airtight container.

SUNSHINE GRANOLA

Yield: 12 cups

3 cups regular oatmeal,
  uncooked
1 (1½ oz.) package sesame
  seeds
1 cup sunflower seeds
1 cup unsweetened wheat germ
½ cup vegetable oil

½ cup honey
1 cup golden seedless raisins
1 cup diced dried apricots
1 cup chopped dates
1 cup flaked coconut
1 cup sliced almonds

Preheat oven to 250°.   Cookie sheet

Lightly grease cookie sheet. In a large bowl, combine oatmeal, sesame seeds, sunflower seeds, and wheat germ. Stir oil and honey together and pour over dry mixture, stirring well. Spread mixture on cookie sheet and bake for 45 minutes. Allow mixture to cool and then break into large pieces. Combine pieces with raisins, apricots, dates, coconut, and almonds. Store in an airtight container.

Can be stored in an airtight container for up to two months in refrigerator.

GUIDE TO CANDY MAKING

The two methods of assuring candy has been cooked to the proper consistency are:

1) the use of a candy thermometer to record degrees
2) the cold water test

To use the cold water test, a fresh cup of chilled water must be used for each test. Remove candy from heat and drop about ½ teaspoon candy into the chilled water. If possible, pick the candy up and roll into a ball.

SOFT BALL TEST (234°-238°): the candy rolled into a soft ball will quickly lose its shape when removed from water. This applies to fondant and fudge.

FIRM BALL TEST (245°-248°): the candy rolled into a firm, but not hard ball, will flatten shortly after being removed from water. This applies to divinity and caramels.

HARD BALL TEST (265°-270°): the candy rolled into a hard ball will roll on a plate when removed from water. This applies to taffy.

SOFT CRACK TEST (275°-280°): the candy will form brittle threads that will soften on removal from water. This applies to butterscotch.

HARD CRACK TEST (285°-290°): the candy will form brittle threads in the water that will remain hard upon removal. This applies to peanut brittle.

CARAMELIZED SUGAR TEST (310°-321°): the sugar will melt, then turn to a golden brown color, and will form a hard brittle ball in chilled water.

SUBSTITUTIONS

If a recipe calls for:	Substitute:
1 teaspoon allspice	½ teaspoon cinnamon plus ⅛ teaspoon cloves plus ¼ teaspoon nutmeg
1 garlic clove	½ teaspoon garlic powder
1 tablespoon fresh grated ginger	1 teaspoon dry ginger
1 teaspoon dry mustard	1 tablespoon prepared mustard
1 tablespoon fresh herbs	1 teaspoon dry herbs
1 teaspoon poultry seasoning	¼ teaspoon thyme plus ¾ teaspoon sage
1 teaspoon pumpkin pie spice	½ teaspoon cinnamon plus ½ teaspoon ginger plus ½ teaspoon allspice plus ⅛ teaspoon nutmeg
1 cup all-purpose flour	1 cup whole wheat flour minus 2 tablespoons
1 cup cake flour, sifted	1 cup all-purpose flour minus 2 tablespoons
1 cup self-rising flour	1 cup all-purpose flour plus 1½ teaspoons baking powder
1 teaspoon baking powder	¼ teaspoon soda plus ½ teaspoon cream of tartar
1 tablespoon cornstarch	2 tablespoons all-purpose flour (for thickening) or 4 teaspoons tapioca
3 cups dry cornflakes	1 cup crushed cornflakes

If a recipe calls for:	Substitute:
1 cup powdered sugar	1 cup granulated sugar plus 1 tablespoon cornstarch, blended on medium high for 2 minutes
1 cup granulated sugar	¾ cup honey or 1 cup unsweetened applesauce or ½ cup unsweetened frozen fruit juice concentrate
1 cup honey	¾ cup granulated sugar plus ¼ cup liquid
1 cup butter	⅞ cup corn oil (or shortening) plus ½ teaspoon salt
1 cup buttermilk	¾ cup milk 1 tablespoon white vinegar, stir and let thicken five minutes
1 whole egg	2 egg whites
1 cup heavy cream	1 cup evaporated skim milk
1 cup sour cream	1 cup evaporated milk or 1 cup heavy cream plus 1 tablespoon vinegar
1 cup skim milk	⅓ cup nonfat dry milk powder plus ¾ cup water
1 ounce semi-sweet chocolate	3 tablespoons unsweetened cocoa powder plus 2 tablespoons water plus 1 teaspoon honey
1 square unsweetened chocolate	3 tablespoons cocoa powder plus 1 tablespoon butter

If a recipe calls for:	Substitute:
3 medium bananas	1 cup mashed bananas
1 medium lemon	2-3 tablespoons juice or 1½ teaspoons lemon flavoring
10 miniature marshmallows	1 large marshmallow
8 ounces fresh mushrooms	6 ounces canned mushrooms
1 medium onion	2 tablespoons instant (chopped or minced) onion flakes or 1½ teaspoons onion powder
1 cup canned tomatoes	1⅓ cups chopped fresh tomatoes, simmered for 10 minutes
1 cup tomato ketchup	1 cup tomato sauce plus ½ cup sugar plus 2 tablespoons vinegar
1 cup tomato sauce	½ cup tomato paste plus ¾ cup water
1 teaspoon Worcestershire sauce	1 tablespoon soy sauce plus dash of cayenne pepper
1 cup margarine	¾ cup vegetable oil

PAN SUBSTITUTIONS

If a recipe calls for:	Substitute:
1 (10" x 4") tube	1 (13" x 9") rectangle or 2 (15" x 10") rectangles or 2 (9" x 5") loafs
1 (8" x 4") loaf	1 (8" x 8") square

If a recipe calls for:

Substitute:

If a recipe calls for:	Substitute:
1 (12" x 8") rectangle	2 (8 inch) layers
1 (13" x 9") rectangle	2 (8" x 8") squares or 2 (9 inch) layers
1 (9" x 5") loaf	24 to 30 (2½ inch) cupcakes or 1 (9" x 9") square
1 (9" x 3½") tube	2 (9 inch) layers or 24 to 30 (2½ inch) cupcakes
1 (9" x 9") square	2 thin (8 inch) layers
2 (9" x 9") squares	3 (8 inch) layers
2 (8" x 8") squares	2 (9 inch) layers or 1 (13" x 9") rectangle
1 (8" x 8") square	1 (9 inch) layer
2 (9 inch) layers	1 (15" x 10") rectangle or 30 (2½ inch) cupcakes or 2 (8" x 8") squares or 3 thin (8 inch) layers
3 (8 inch) layers	2 (9" x 9") squares
2 (8 inch) layers	2 thin (8" x 8") squares or 18 to 24 (2½ inch) cupcakes

SUN-SATIONAL SACK SNACKS

❋ Border Twirls
Layer Cheddar cheese spread, lettuce leaf, and a thin slice of ham, roast beef, or turkey on a flour tortilla (page 39). Roll up and cut into three pieces.

❋ Tex-Mex Pita
Fill whole wheat pita pocket with leftover Fajita slices (page 40), shredded lettuce, Cheddar, and tomato chunks. Drizzle with a mild taco sauce.

❋ Fiesta Feast
Stuff Black Bean Salad (page 37) and shredded roast chicken into mini white or wheat pita pockets.

❋ Chunky Monkey
Mix cream cheese with chopped pecans, apple chunks, nutmeg, and cinnamon, then spread on Banana Bread (page 121).

❋ Tropical Cheese Kabobs
Dip chunks of fresh fruit in orange juice then alternate with cheese chunks on a wooden skewer. Snip off sharp ends of skewer.

❋ Nutmeg & Honey Dip
Beat equal parts softened margarine and honey. Sprinkle with a little nutmeg. Dip cubed French Bread (page 115). This is a perfect substitute for a butter and honey sandwich.

❋ PB & Ap-sicles
Put an apple on a popsicle stick, coat with peanut butter, and roll in Granola (page 272). Wrap in wax paper.

❋ Nuts about Pasta
Toss pasta with chopped vegetables, peanuts, and vinaigrette (page 94).

❋ Itty-Bitty Bites
Cut tiny Buttermilk Biscuits (page 121) in half, spread with honey mustard, and layer with bite-size slices of ham or turkey.

❋ Appleseed's Favorite
Spread apple butter on Whole Wheat Rolls (page 118), top with grated Cheddar cheese.

❋ The Double Dip
Mix ½ cup ricotta cheese with 1 tablespoon strawberry jam as a dip or use lemon or vanilla yogurt as a dip. For the second dip, combine ¼ cup granola cereal (page 272) with 1 tablespoon colored sprinkles. Dip any fruit or try toasted raisin breadsticks or waffle strips.

❋ Club Bagel
Slice a bagel into thirds and layer with bagel on bottom, chicken salad (page 105), three slices of bacon, bagel in middle, mayo, Muenster cheese, bagel on top.

❋ Any dressings or sauces may be packed in separate container and added when time to eat.

❋ A frozen juice box will keep a lunch box cold and all food fresh until lunch time. You then have an extra cold and sometimes slushy drink, too. This even keeps yogurt and jello cold!

ORDER FORM

Mail to:

Junior League of McAllen, Inc.
P.O. Box 2465
McAllen, TX 78502-2465

Please send _____ copy(ies) Some Like It Hot   @ $16.95* each $_____
              _____ copy(ies) *La Piñata*   @ $15.95* each $_____
         OR both books for $29.90 (a $3.00 savings) $_____
Postage & handling   @ $3.00* each $_____
Gift wrapping (includes sales tax)   @ $2.15* each $_____
                    *Prices subject to change without notice
                                                        TOTAL $_____

Name_____
Address_____
City_____State_____Zip_____
              Make checks payable to *Junior League of McAllen, Inc.*

ORDER FORM

Mail to:

Junior League of McAllen, Inc.
P.O. Box 2465
McAllen, TX 78502-2465

Please send _____ copy(ies) Some Like It Hot   @ $16.95* each $_____
              _____ copy(ies) *La Piñata*   @ $15.95* each $_____
         OR both books for $29.90 (a $3.00 savings) $_____
Postage & handling   @ $3.00* each $_____
Gift wrapping (includes sales tax)   @ $2.15* each $_____
                    *Prices subject to change without notice
                                                        TOTAL $_____

Name_____
Address_____
City_____State_____Zip_____
              Make checks payable to *Junior League of McAllen, Inc.*

ORDER FORM

Mail to:

Junior League of McAllen, Inc.
P.O. Box 2465
McAllen, TX 78502-2465

Please send _____ copy(ies) Some Like It Hot   @ $16.95* each $_____
              _____ copy(ies) *La Piñata*   @ $15.95* each $_____
         OR both books for $29.90 (a $3.00 savings) $_____
Postage & handling   @ $3.00* each $_____
Gift wrapping (includes sales tax)   @ $2.15* each $_____
                    *Prices subject to change without notice
                                                        TOTAL $_____

Name_____
Address_____
City_____State_____Zip_____
              Make checks payable to *Junior League of McAllen, Inc.*

Where did you hear about this cookbook? _____

_____

What local stores would you like to see carry this cookbook? _____

_____

Store Name _____ Phone ( ) _____
Address _____
City _____ State _____ Zip _____
Was book purchased as a gift? _____
What attracted you to this particular book? _____
What is your age? _____

- -

Where did you hear about this cookbook? _____

_____

What local stores would you like to see carry this cookbook? _____

_____

Store Name _____ Phone ( ) _____
Address _____
City _____ State _____ Zip _____
Was book purchased as a gift? _____
What attracted you to this particular book? _____
What is your age? _____

- -

Where did you hear about this cookbook? _____

_____

What local stores would you like to see carry this cookbook? _____

_____

Store Name _____ Phone ( ) _____
Address _____
City _____ State _____ Zip _____
Was book purchased as a gift? _____
What attracted you to this particular book? _____
What is your age? _____